Antibacterial Therapy in Adults with COVID-19

Antibacterial Therapy in Adults with COVID-19

Editors

Sabine Danielle Allard
Johan Van Laethem

MDPI • Basel • Beijing • Wuhan • Barcelona • Belgrade • Manchester • Tokyo • Cluj • Tianjin

Editors
Sabine Danielle Allard
Internal Medicine
Vrije Universiteit Brussel
Brussels
Belgium

Johan Van Laethem
Internal Medicine
Vrije Universiteit Brussel
Brussels
Belgium

Editorial Office
MDPI
St. Alban-Anlage 66
4052 Basel, Switzerland

This is a reprint of articles from the Special Issue published online in the open access journal *Antibiotics* (ISSN 2079-6382) (available at: www.mdpi.com/journal/antibiotics/special_issues/Antibacterial_COVID19).

For citation purposes, cite each article independently as indicated on the article page online and as indicated below:

LastName, A.A.; LastName, B.B.; LastName, C.C. Article Title. *Journal Name* **Year**, *Volume Number*, Page Range.

ISBN 978-3-0365-6813-3 (Hbk)
ISBN 978-3-0365-6812-6 (PDF)

Contents

Preface to "Antibacterial Therapy in Adults with COVID-19"

Since the emergence of the Coronavirus-19 infectious disease (COVID-19) pandemic, a rising number of reports have underlined the risk of increasing antimicrobial resistance due to antibiotic overuse in COVID-19 patients. In addition, many physicians and pharmacists involved in antimicrobial stewardship have had to shift their activities to the containment of the COVID-19 crisis. In contrast to this observed overconsumption of antibiotics, very few bacterial superinfections have been documented in COVID-19 patients, especially in patients admitted outside the intensive care unit and in the first days of admission. However, the identification of bacterial co-/superinfections in COVID-19 patients is difficult, as inflammatory and radiological markers of bacterial infection lack specificity in this setting. Furthermore, studies regarding the effect of immune suppression, including the use of corticosteroids and anti-interleukins, and the effect of potential immunomodulatory properties of certain antibiotics on the occurrence of bacterial co-/superinfection are needed.

This Special Issue of *Antibiotics* aims to increase our knowledge regarding (more or less specific) markers associated with bacterial co-/superinfection in COVID-19 patients, quantitative and qualitative data regarding antibiotic prescriptions in COVID-19 patients, potential beneficial effects of antibiotic use in certain COVID-19 subgroups, detrimental effects associated with antibiotic overuse in COVID-19 patients, and evidence-based guidelines, which could facilitate the decision-making process when antibiotic prescriptions are considered.

Sabine Danielle Allard and Johan Van Laethem
Editors

antibiotics

MDPI

Article

Effect of Tocilizumab on Mortality in Patients with SARS-CoV-2 Pneumonia Caused by Delta or Omicron Variants: A Propensity-Matched Analysis in Nimes University Hospital, France

Paul Laffont-Lozes [1,2], Didier Laureillard [2], Paul Loubet [2], Robin Stephan [3], Myriam Chiaruzzi [2], Edouard Clemmer [2], Aurelie Martin [2], Claire Roger [4], Laurent Muller [4], Pierre-Géraud Claret [5], Radjiv Goulabchand [6], Clarisse Roux [1], Jean-Philippe Lavigne [3,7], Albert Sotto [2,7] and Romaric Larcher [2,8,*]

1 Department of Pharmacy, Nimes University Hospital, Place du Professeur Robert Debre, 30000 Nimes, France
2 Infectious and Tropical Diseases Department, Nimes University Hospital, Place du Professeur Robert Debre, 30000 Nimes, France
3 Department of Microbiology and Hospital Hygiene, Nimes University Hospital, Place du Professeur Robert Debre, 30000 Nimes, France
4 Anesthesiology and Critical Care Medicine Department, Nimes University Hospital, Place du Professeur Robert Debre, 30000 Nimes, France
5 Emergency Medicine Department, Nimes University Hospital, Place du Professeur Robert Debre, 30000 Nimes, France
6 Department of Internal Medicine, Nimes University Hospital, Place du Professeur Robert Debre, 30000 Nimes, France
7 VBIC (Bacterial Virulence and Chronic Infection), INSERM (French Institute of Health and Medical Research), Montpellier University, 34090 Montpellier, France
8 PhyMedExp (Physiology and Experimental Medicine), INSERM (French Institute of Health and Medical Research), CNRS (French National Centre for Scientific Research), University of Montpellier, 34090 Montpellier, France
* Correspondence: romaric.larcher@chu-nimes.fr; Tel.: +33-646-668-4149

Citation: Laffont-Lozes, P.; Laureillard, D.; Loubet, P.; Stephan, R.; Chiaruzzi, M.; Clemmer, E.; Martin, A.; Roger, C.; Muller, L.; Claret, P.-G.; et al. Effect of Tocilizumab on Mortality in Patients with SARS-CoV-2 Pneumonia Caused by Delta or Omicron Variants: A Propensity-Matched Analysis in Nimes University Hospital, France. *Antibiotics* **2023**, *12*, 88. https://doi.org/10.3390/antibiotics12010088

Academic Editors: Sabine Danielle Allard and Johan Van Laethem

Received: 23 November 2022
Revised: 25 December 2022
Accepted: 30 December 2022
Published: 4 January 2023

Abstract: We aimed to assess the factors associated with mortality in patients treated with tocilizumab for a SARS-CoV-2 pneumonia due to the delta or omicron variants of concern (VOC) and detect an effect of tocilizumab on mortality. We conducted a prospective cohort study in a tertiary hospital from 1 August 2021 to 31 March 2022 including patients with severe COVID-19, treated with tocilizumab. Factors associated with mortality were assessed in a Cox model; then, the 60-day mortality rates of COVID-19 patients treated with standard of care (SoC) +/− tocilizumab were compared after 1:1 propensity score matching. The mortality rate was 22% (N = 26/118) and was similar between delta and omicron cases (p = 0.6). The factors independently associated with mortality were age (HR 1.06; 95% CI (1.02–1.11), p = 0.002), Charlson index (HR 1.33; 95% CI (1.11–1.6), p = 0.002), WHO-CPS (HR 2.56; 95% CI (1.07–6.22) p = 0.03), and tocilizumab infusion within the first 48 h following hospital admission (HR 0.37, 95% CI (0.14–0.97), p = 0.04). No significant differences in mortality between the tocilizumab plus SoC and SoC alone groups (p = 0.5) were highlighted. However, the patients treated with tocilizumab within the 48 h following hospital admission had better survival (p = 0.04). In conclusion, our results suggested a protective effect on mortality of the early administration of tocilizumab in patients with severe COVID-19 regardless of the VOC involved.

Keywords: COVID-19; variant of concern; interleukin-6 receptor antagonist; early administration; mortality rates

1. Introduction

Soon after the beginning of the pandemic, the pathogenicity of the severe acute respiratory syndrome coronavirus 2 (SARS-CoV-2) was reported to be related to a "Cytokine storm" [1]. This deregulation in cytokine secretion was mainly characterized by an extensive expression of interleukin-6 (IL-6) and tumor necrosis factor α (TNF-α) which injure the lungs and other organs to a lesser extent [1,2]. Unsurprisingly, corticosteroids were the first immunomodulatory medications to demonstrate a mortality benefit among patients with coronavirus disease 2019 (COVID-19) and are now a cornerstone of COVID-19 treatment in patients receiving respiratory support [3].

Early on, the potential benefit of IL-6 receptor antagonist monoclonal antibodies such as tocilizumab was suggested [2]. Indeed, since IL-6 was identified as one of the key cytokines involved in the COVID-19-induced cytokine storm, tocilizumab has been proposed to block the downstream signal transduction by binding IL-6 receptors. Thus, it blocks the JAK/STAT tyrosine kinase system and the Ras/mitogen activated protein kinase (MAPK)/NF-κB-IL-6 pathway, thereby limiting the cytokine storm and its life-threatening consequences [4].

In mid-2021, two large platform trials reported that tocilizumab reduced mortality and the use of invasive mechanical ventilation, and increased the chances of a successful hospital discharge [5,6]. Subsequently, numerous studies have investigated its effect in patients with SARS-CoV-2 pneumonia caused by the alpha, beta or gamma variants of concern (VOC) [7–11], even recently reporting that the antagonist of IL-6 receptors also improved long-term outcomes [12]. Unfortunately, in numerous countries, several factors have limited the use of tocilizumab, such as drug shortages, high cost or late approval (December 2021 in Europe, December 2022 in the USA). Therefore, real-life data on tocilizumab treatment remain scarce, especially in patients infected with delta or omicron VOCs [13–15]. Moreover, its use is questionable in omicron cases. Indeed, the reduced intrinsic virulence of the omicron VOC and immunization might contribute to decrease the inflammatory response, reducing the COVID-19 severity, and accordingly reducing the benefit of tocilizumab treatment [16].

This study aimed to report real-life data on tocilizumab use in patients with SARS-CoV-2 pneumonia caused by the delta or omicron VOC. The main objective was to assess factors associated with mortality in patients treated with tocilizumab plus standard of care (SoC). Then, we aimed to compare the prognosis of patients treated with tocilizumab plus SoC to those treated with SoC alone, to detect a potential effect of tocilizumab on mortality.

2. Results

2.1. Population

Characteristics of the study population are reported in Table 1.

Table 1. Patients' characteristic and outcomes.

Characteristics	Total (N = 118)	Survivor (N = 92)	Non-Survivor (N = 26)	*p*-Value
Age, years	69 (56–77)	64 (54–72)	78 (71–84)	<0.001 *
Male	79 (67%)	61 (66%)	18 (69%)	0.78
BMI, kg/m^2 [a]	28 (25–31)	28 (25–31)	27 (24–31)	0.20
Main comorbidities				
Diabetes	36 (31%)	24 (26%)	12 (46%)	0.05
Myocardial ischemia	18 (15%)	9 (10%)	9 (35%)	0.003 *
Chronic heart failure	5 (4%)	2 (2%)	3 (12%)	0.06
Chronic lung disease	18 (15%)	12 (13%)	6 (23%)	0.22
Chronic kidney disease	7 (6%)	4 (4%)	3 (12%)	0.19
Dementia	5 (4%)	1 (1%)	4 (15%)	0.01 *
Cancer	19 (16%)	11 (12%)	8 (31%)	0.03 *
Hemopathy [b]	8 (7%)	4 (4%)	4 (15%)	0.06
Charlson index	1 (0–2)	1 (0–2)	2.5 (1–5)	<0.001 *
Chest CT-Scan [c]				
Mild	13 (11%)	8 (9%)	5 (19%)	0.14
Moderate	72 (61%)	59 (64%)	13 (50%)	0.2
Severe	24 (20%)	18 (20%)	6 (23%)	0.7
Critical	8 (7%)	6 (7%)	2 (8%)	0.83

Table 1. *Cont.*

Characteristics	Total (N = 118)	Survivor (N = 92)	Non-Survivor (N = 26)	*p*-Value
Variant of concern				
Delta	101 (86%)	78 (85%)	23 (88%)	0.64
Omicron	17 (14%)	14 (15%)	3 (12%)	0.64
Oxygen requirement				
Conventional oxygen therapy	43 (36%)	41 (45%)	2 (8%)	0.003 *
High-flow nasal canula	57 (48%)	39 (42%)	18 (69%)	0.02 *
Mechanical ventilation	18 (15%)	12 (13%)	6 (23%)	0.22
ECMO [d]	6 (5%)	5 (5%)	1 (4%)	0.75
WHO-CPS [e] at admission	5 (5–6)	5 (5–6)	5 (5–6)	0.66
WHO-CPS at tocilizumab administration	6 (5–6)	5 (5–6)	6 (5–6)	0.047 *
Tocilizumab treatment				
Administration timing, days	2 (2–4)	2 (2–3)	3.5 (2–5)	0.002 *
Administration within 2 days	68 (58%)	59 (64%)	9 (35%)	0.009 *
Inflammation biomarker				
CRP pre-tocilizumab, mg/L	124 (89–165)	128 (91–166)	103 (77–162)	0.45
CRP post-tocilizumab, mg/L	47 (23–89)	47 (23–81)	54 (24–95)	0.25
Outcomes				
Death at day 60	26 (22%)	-	-	
ICU [f] admission	56 (47%)	39 (42%)	17 (65%)	0.04 *
Limitation of life support	20 (17%)		20 (77%)	
CAPA [g]	3 (3%)	1 (<1%)	2 (8%)	0.1

Results are expressed as median and interquartile range (IQR) or as number of patients and percentage (%) as appropriate. [a] BMI: body mass index, [b] Hemopathy: lymphoma, leukemia or myeloma. [c] Chest CT-Scan were classified according to percentage of pulmonary involvement: Mild (10–25%); Moderate (25–50%); severe (50–75%) and critical (75–100%), [d] ECMO: extracorporeal membrane oxygenation, [e] WHO-CPS: World Health Organization Clinical Progression Scale, [f] ICU: intensive care unit, [g] CAPA: COVID-19 associated pulmonary aspergillosis. * Statistically significant.

Among the 1213 patients admitted for COVID-19 during the study period, 118 patients (9.7%) were treated with tocilizumab and included in the study (Figure 1). Of them, 67% were males (n = 79), and the median age and Charlson index were 69 years old (IQR, 56–77) and 1 (IQR, 0–2), respectively.

At hospital admission, the World Health Organization Clinical Progression Scale (WHO-CPS) was 5 in 68 patients (58%) and 6 in 50 patients (42%). The median C-reactive protein (CRP) level was 124 mg/L (IQR, 89–165), and the pulmonary involvement on CT-Scan was classified as moderate or severe in more than 80% of the patients. A total of 101 patients (86%) were infected with delta, and 17 (14%) with omicron.

Tocilizumab was administered at a median dose of 600 mg (IQR, 600–800), and 68 patients (58%) received their first tocilizumab dose within the first 48 h of hospital stay. Only six patients (5%) had a second dose.

2.2. Outcomes

The median CRP decreased to 47 mg/L (IQR, 23–89) following the tocilizumab infusion. The decrease in CRP was shown in all patients but 9, and the WHO-CPS increased in 34 patients (29%) (Figure 2).

The median length of hospital stay was 10 days (IQR, 7–15). At the end of the follow-up period of 60 days, 88 patients (75%) were discharge alive (WHO-CPS $\leq$ 3), 4 patients (3%) were still hospitalized with a WHO-CPS score of 4, 5, 6 and 9, respectively, and 26 patients (22%) had died (Table 1).

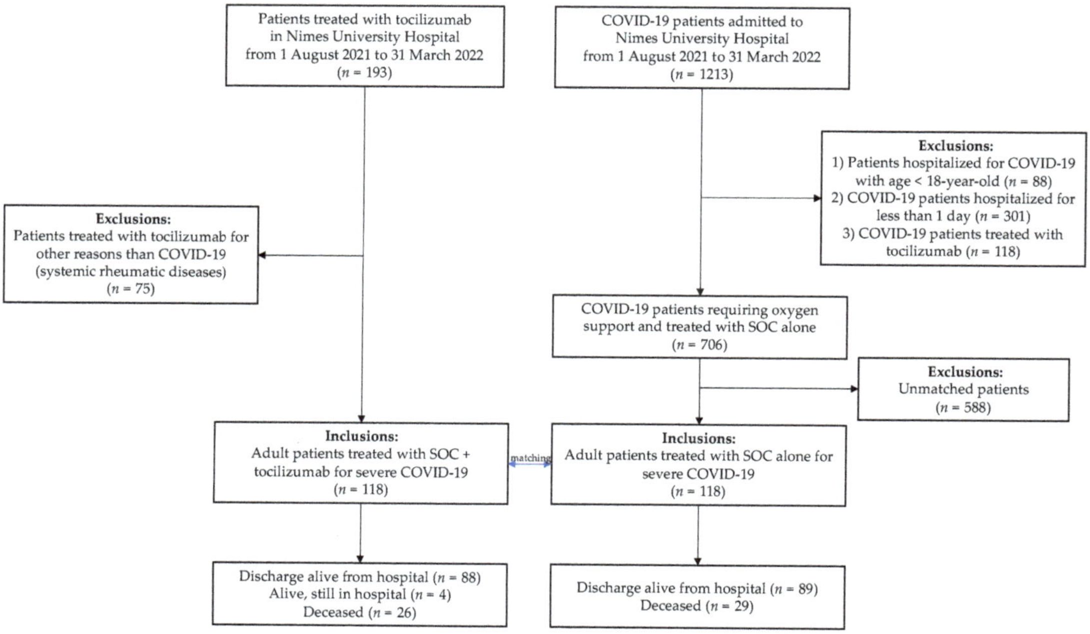

Figure 1. Flow chart of the study population, depicting the selection of patients treated with tocilizumab for severe COVID-19 and the selection of patients with COVID-19 not treated with tocilizumab (matched control group on age, sex, oxygen needs, Charlson index and variant of concern).

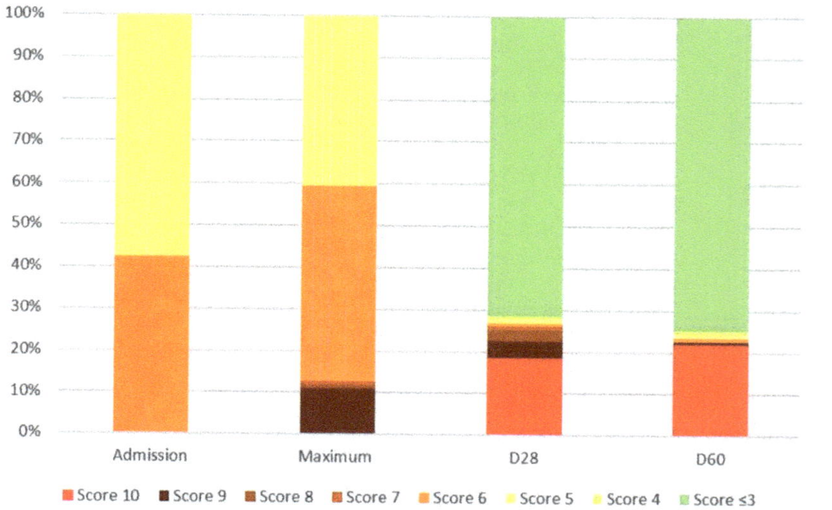

Figure 2. World Health Organization Clinical Progression Scale (WHO-CPS) evolution during the 60-day follow-up. (D28 and D60: assessment of WHO-CPS 28 and 60 days after admission).

Importantly, the 60-day survival probability was at 87%, 95% CI (79–95%) in patients treated with tocilizumab within the first 48 h following hospital admission, and at 66%, 95% CI (54–80%) in those treated later ($p = 0.009$). Moreover, the 60-day survival probability was not statistically different in delta and omicron cases (77%, 95% CI (69–86%) versus 82%, 95% CI (66–100%), $p = 0.6$), as illustrated in Figure 3.

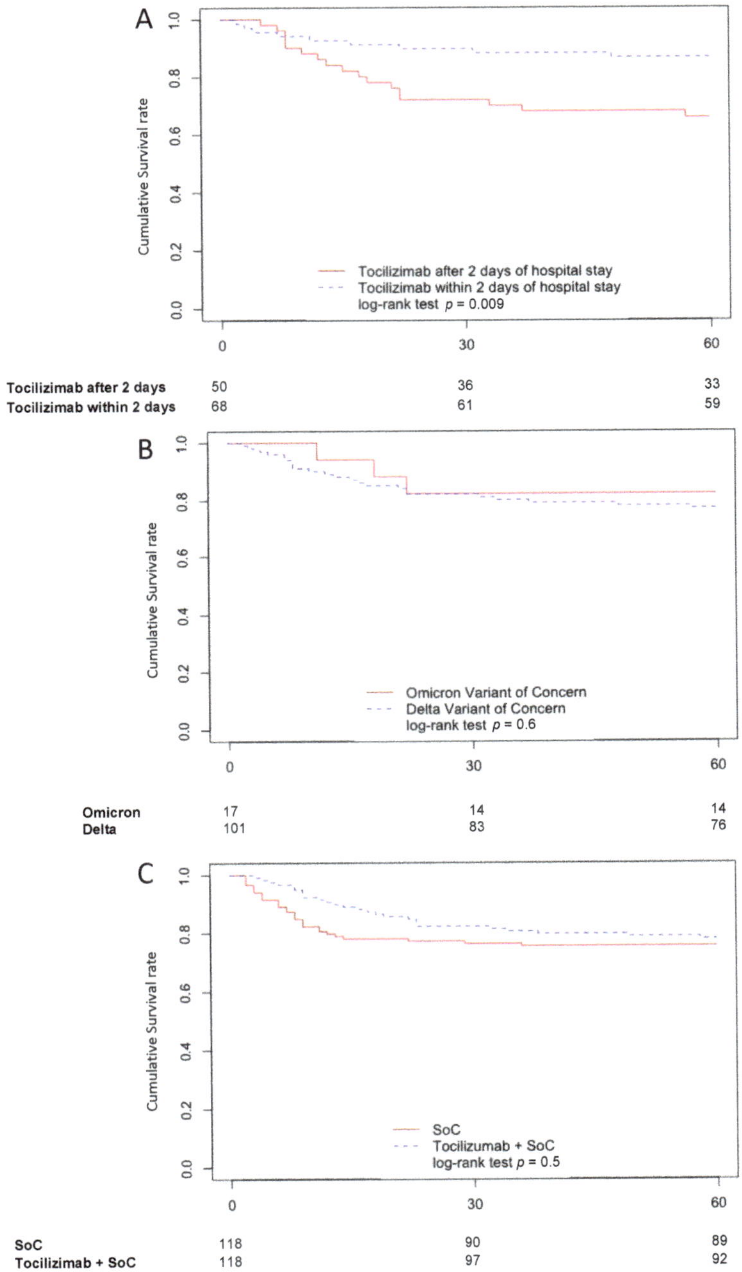

Figure 3. *Cont.*

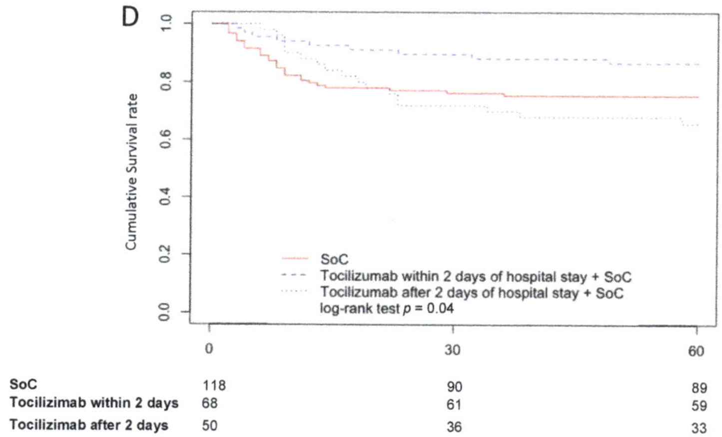

Figure 3. Kaplan–Meier Survival Curves at day 60 for patients treated with tocilizumab within 2 days of hospital admission and patients treated with tocilizumab after 2 days (**A**), in omicron and delta cases treated with tocilizumab (**B**), in patients treated with standard of care (SoC) alone and with SoC plus tocilizumab (**C**), in patient treated with SoC alone and patients treated with tocilizumab plus SoC within 2 days of hospital admission and patients treated with tocilizumab plus SoC after 2 days (**D**).

2.3. *Factors Associated with Mortality*

Results of the univariate and multivariable Cox regressions are presented in Table 2.

Table 2. Factors associated with mortality in patients treated with tocilizumab.

	Univariable Analysis			Multivariable Analysis		
Characteristics	**HR**	**95% CI**	***p*-Value**	**HR**	**95% CI**	***p*-Value**
Male	1.2	0.5–2.7	0.73			
Age	1.1	1–1.1	<0.001 *	1.06	1.02–1.11	0.002 *
BMI	0.95	0.88–1	0.2 *	0.99	0.91–1.08	0.83
Charlson index	1.5	1.3–1.7	<0.001 *	1.33	1.11–1.6	0.002 *
Chest CT-Scan [a]						
Mild	2.1	0.78–5.5	0.14 *	0.99	0.31–3.12	0.98
Moderate	0.61	0.28–1.3	0.21			
Severe	1.2	0.47–2.9	0.74			
Critical	1.3	0.3–5.3	0.76			
Variant of concern						
Delta	1.3	0.4–4.4	0.65			
Omicron	0.76	0.23–2.5	0.65			
WHO-CPS [b]						
At admission	1.2	0.54–2.5	0.69			
At tocilizumab administration	2.3	0.98–5.2	0.055 *	2.58	1.07–6.22	0.03 *
Inflammation biomarkers						
CRP pre-tocilizumab	1	0.99–1	0.49			
CRP post-tocilizumab	1	1–1	0.16 *	1	1–1.01	0.27
Tocilizumab treatment						
Administration within 2 days	0.35	0.16–0.73	0.012 *	0.37	0.14–0.97	0.04 *
Dose	1	0.99–1	0.16 *	1	0.99–1	0.57

[a] Chest CT-Scans were classified according to percentage of pulmonary involvement: Mild (10–25%); Moderate (25–50%); severe (50–75%) and critical (75–100%), [b] WHO-CPS: World Health Organization Clinical Progression Scale. * Statistically significant.

In the multivariable analysis, age (HR 1.06; 95% CI (1.02–1.11), $p = 0.002$), Charlson index (HR 1.33; 95% CI (1.11–1.60), $p = 0.002$) and WHO-CPS score at the time of tocilizumab infusion (HR 2.58; 95% CI (1.15–1.66), $p = 0.03$) were independently associated with mortality. In contrast, early administration of tocilizumab had a protective effect (HR 0.37; 95% CI (0.14–0.97), $p = 0.04$).

2.4. Tocilizumab Treatment Effect on Mortality

Using propensity score matching in a Cox model, tocilizumab treatment was shown to have no effect on mortality ($p = 0.5$). However, as reported in Figure 3, the 60-day survival probability was higher in patients treated with tocilizumab within the first 48 h of their hospital stay compared to those treated later and those treated with SoC alone (87%, 95% CI (79–95%) vs. 66%, 95% CI (54–80%) vs. 75%, 95% CI (68–84%), $p = 0.04$).

In addition, when we included variables resulting in an imbalance between the groups after propensity score matching (namely, age, high-flow oxygen support needs and mechanical ventilation needs), the effect of tocilizumab was significant in the adjusted Cox model (HR 0.48, 95% CI (0.27–0.86), $p = 0.01$).

Of note, COVID-19 associated pulmonary aspergillosis (CAPA) was diagnosed in 3 patients among those patients treated with tocilizumab (3/118; 3%) and in 6 among those treated with SoC alone (6/706; 1%). All of them required mechanical ventilation.

3. Discussion

This study reported the results of a single-center cohort that included 118 patients with severe COVID-19 treated with tocilizumab on top of SoC and described their clinical features and outcomes in a period of the circulation of the delta and omicron variants. We found a 22% overall 60-day mortality rate, similar in both omicron and delta cases. A higher age, higher Charlson index and higher level of oxygen support were independently associated with mortality. In contrast, the early administration of tocilizumab was independently associated with better survival. Importantly, after propensity score matching, this study did not find a significant effect for tocilizumab + SoC on 60-day mortality compared to SoC alone. However, the effect of tocilizumab was significant after adjustment for variables resulting in an imbalance between the groups.

In our cohort of patients treated with tocilizumab, as has been widely reported in numerous previous studies on COVID-19 patients [17], mortality was independently associated with higher age and a higher Charlson index. Among comorbidities, diabetes has been reported as a critical risk factor for severe SARS-CoV-2 infection and death [18], as highlighted by our result (death rates of 43% vs. 23%). Similarly, the level of hypoxemia assessed by PaO_2/FiO_2, the SpO_2/FiO_2 ratio or the WHO-CPS is one of the most significant factors previously reported to drive mortality in COVID-19 patients [10].

Other studies [7,8,10,11,19], including platform trials [5,6], have reported a favorable effect of tocilizumab in alpha, beta or gamma cases; however, the benefit of tocilizumab in patients infected with delta, which is reported to be more virulent [20], or in patients infected with omicron, which is, in contrast, reputed to be less virulent [21], remained disputable. To the best of our knowledge, this study was the first to report the use of tocilizumab in a large cohort of severe COVID-19 patients infected with the delta and omicron VOC, suggesting that, among the patients developing a severe form of the disease, the mortality rates were similar in the omicron and delta cases, as it has been reported with previous VOCs [14]. In the same line, some authors have highlighted that the rate of severe COVID-19 may change depending on the VOC, but, in cases of severe forms, they reported that the cytokine secretion remained stable and independent from the VOC [22]. Accordingly, our results suggested that tocilizumab remained an efficient option for the treatment of severe COVID-19 patients in association with corticosteroids, regardless of the VOC involved.

One striking finding of this study was that early administration of tocilizumab (within 48 h following hospital admission) was associated with a better outcome. This result was in accordance with those reported in other studies suggesting the benefit of tocilizumab administration within 2–3 days of hospital admission [8,10,11]. Similarly, the results of the sensitivity analyses of RECOVERY trials [5] have also suggested that patients who received tocilizumab within seven days of symptom onset are those who benefited the most. It is worth noting that some authors [23] have highlighted that tocilizumab's effect on mortality is more related to its rapid administration after the onset of oxygen requirement than after symptom onset. Accordingly, the probability of any benefit of tocilizumab has been estimated at 99%, 96% and 75%, for patient receiving simple oxygen, non-invasive ventilation and mechanical ventilation, respectively [23]. Our results, consistent with these studies, suggest that administering tocilizumab as soon as a COVID-19 patient requires oxygen support may provide the greatest survival benefit.

However, after two years of the pandemic, tocilizumab remains a debated therapeutic option for some physicians caring for COVID-19 patients. First, conflicting results [24–26] discouraged some of them from using it. Second, tocilizumab has been reported to be associated with an increased risk of infections [27], and especially an increased risk of CAPA [28]. However, COVID-19 itself and corticosteroids also increased this risk [29,30], whereas few superinfections, especially CAPA, have been reported in large trials [5,6]. In accordance with these reports, during the study period, only 12 COVID-19 patients were diagnosed with proven, probable or possible CAPA [31] in our center, and among them only 3 had been treated with tocilizumab. Moreover, despite this possible increased risk of health-care associated infections, the benefits of tocilizumab use seem to outweigh the risks in severe COVID-19 patients, since it decreased all-cause mortality in the short [5,6] and long term [12].

Some data have suggested a second dose of tocilizumab in the case of weak clinical improvement [5]. However, the patients who could benefit from a second dose remain poorly identified. In our cohort, 6 patients who had increased their oxygen requirements 12–24 h after a first dose of tocilizumab subsequently received a second dose. All of them were discharged alive. Moreover, in our study, among the 9 patients whose CRP increased after a first infusion of tocilizumab, none received a second dose, and 4 (44%) of them died. Along the same lines, Khurshid et al. reported that the COVID-19 patients who maintained higher CRP values during treatment had the worst outcomes [32]. However, further exploration is mandatory to assess the use of biomarkers, such as CRP, to trigger a second dose of tocilizumab.

Our study has both strengths and limitations. First, our conclusions are limited by the study's monocentric design, which could induce bias in the interpretation of the results and limit their generalization. However, this bias is limited by the standardization of care for COVID-19 patients in accordance to the WHO international guidelines [33]. Moreover, all patients received dexamethasone, which may synergistically interact with tocilizumab [34]. Second, the data were retrospectively collected for our control group, which could induce bias in data collection. Third, we performed a propensity score analysis and found that tocilizumab had no effect on mortality. However, the relatively small size of the cohort could have limited the detection of its effect on mortality. Indeed, based on previously published data [5,6], we should have included 3902 patients in the study (alpha = 0.05 and power 90%) to detect a 4% reduction in mortality. Finally, we reported mortality rates similar to those of randomized clinical trials [5,6,12]; however, in our cohort, the patients were older [35] and had more severe cases (according to our institutional guidelines, tocilizumab was given to those with rapid increase in oxygen requirements), which could have increased mortality rates.

4. Materials and Methods

4.1. Study Design and Settings

This prospective, monocentric, observational cohort study was carried out in a French University Hospital from 1 August 2021 to 31 March 2022.

In our hospital, during the study period, the bed capacity for COVID-19 patients care ranged from 9 to 81 ward-beds and from 41 to 81 ICU-beds.

4.2. Tocilizumab Treatment

Tocilizumab was first approved by the European Medicines Agency (EMA) in 2009 for the treatment of rheumatoid arthritis and juvenile idiopathic arthritis. In June 2021, tocilizumab was granted an emergency use authorization (EUA) for the treatment of COVID-19 in the United States, and the EMA recommended tocilizumab for adults with COVID-19 who are receiving systemic treatment with corticosteroids and require supplemental oxygen or mechanical ventilation in December 2021. More recently, in December 2022, the Food and Drug Administration (FDA) approved tocilizumab for patients with severe COVID-19.

Taking into account the body of evidence available in the literature [5,6], the Anti-Infective Drugs Committee of the Nimes University Hospital authorized the off-label use of tocilizumab in the treatment of severe COVID-19 in June 2021. Thus, tocilizumab started to be administered at physicians' discretion to patients treated with SoC for severe or critical COVID-19 (i.e., intravenous dexamethasone 6mg once a day, prophylactic heparin and oxygen therapy to maintain $SpO_2 > 94\%$). Importantly, remdesivir was not recommended in our center, according to international [33] and national guidelines [36].

The detailed criteria for tocilizumab initiation were as follows: (1) CRP $\geq$ 75 mg/L and an increase of oxygen needs $\geq$ 2L/min within 48 h, or (2) lack of improvement after 48 h of SoC or (3) ICU admission $\leq$ 24 h.

Tocilizumab was administered intravenously at a dose depending on bodyweight: 800 mg if weight > 90 kg; 600 mg if weight > 65 kg and $\leq$90 kg; 400 mg if weight > 40 kg and $\leq$65 kg; and 8 mg/kg if weight $\leq$ 40 kg [5]. A second dose could be given 12–24 h later at the clinician's discretion if the patient's condition had not improved.

Contraindications for tocilizumab were untreated confirmed or suspected bacterial or fungal infection, hepatic cytolysis > 5 upper limit of normal (ULN), neutropenia < 0.5 cells $\times 10^9$/L and thrombopenia < 50 cells $\times 10^9$/L [5].

4.3. Patients

All patients hospitalized during the study period and receiving tocilizumab were screened daily for inclusion using the software of the Pharmacy Department (Pharma®, Computer Engineering, Paris, France). Then, those patients receiving tocilizumab for severe or critical COVID-19 were included in the study. Patients younger than 18 years old or pregnant, those treated with tocilizumab for a reason other than COVID-19 and those with a hospital stay of a length $\leq$ 1 day were excluded (Figure 1). Severe COVID-19 was defined as previously described [37,38].

All consecutive adult COVID-19 patients hospitalized during the study period for severe SARS-CoV-2 pneumonia and treated with SoC were retrieved by screening the hospital database using the International Classification of Diseases [39] to build a control group. Patients younger than 18 years old or pregnant, those treated with tocilizumab and those with a hospital stay of a length $\leq$ 1 day were excluded (Figure 1).

4.4. Data Collection

In patients treated with tocilizumab, their demographical data and morbidities were collected prospectively at hospital admission, and their Charlson index was calculated [40]. The percentage of pulmonary involvement was assessed on a chest CT-Scan at admission and was classified as previously described [41] by a radiologist and confirmed blindly by an infectious disease physician trained in chest CT-Scan interpretation (R.L.). SARS-CoV-2 VOC were also recorded after detection by genotyping or gene sequencing on nasopharyngeal swabs. CRP was recorded 24–48 h before and after tocilizumab infusion. The severity of respiratory failure was assessed using the WHO-CPS at admission and at the time of tocilizumab infusion [42]. In the control group, the demographical data, morbidities, oxygen requirements, VOC and outcomes were collected retrospectively.

4.5. Outcomes

The lengths of the ICU and hospital stays were recorded. The respiratory status was assessed using the WHO-CPS at 28 and 60 days after hospital admission as recommended by the WHO working group on the clinical characterization and management of COVID-19 infection [42]. Vital statuses at ICU and hospital discharge and 60 days after hospital admission were collected; then, mortality rates were calculated. Additionally, cases of CAPA have been retrospectively collected [28].

4.6. Statistical Analysis

Categorical data were described as numbers and percentages, and continuous data as medians with 25th and 75th percentiles (interquartile range: IQR). The population was divided into two groups according to vital status at 60 days. The categorical variables were compared by Chi-square or Fisher's exact test, and the continuous variables were compared by Student's t test or Wilcoxon's rank-sum test as appropriate.

Factors associated with 60-day mortality were assessed using univariable and multivariable cox regression model. Factors associated with 60-day mortality in the univariate analysis (cut-off of $p \leq 0.2$) were included in the multivariable analysis. Proportional hazard assumption was assessed by inspecting the scaled Schoenfeld residuals. Results of Cox regression model were reported as hazard ratio (HR) with 95% confidence interval (95% CI).

Propensity score matching was also performed to compare COVID-19 patients treated with tocilizumab and SoC with those treated with SoC alone. Patients were matched (1:1) with the algorithm for nearest-neighbor matching without replacement, using a maximum tolerance distance between the matched subjects of 0.1 standard deviation. The confounding variables used to calculate the propensity scores were age, BMI ≥ 30 kg/m^2, ICU admission needs, non-invasive high-flow extra-oxygen support needs, mechanical ventilation needs, previous immunosuppressive treatment (including corticosteroids) and previous chemotherapy, as well as each variable included in the Charlson index, its value and the VOC. We identified the variables resulting in an imbalance between the groups after propensity score matching by calculating the standardized mean difference; then, we included these in the subsequent Cox proportional hazards models as covariates to assess the effect of tocilizumab treatment.

Survival curves were obtained using the Kaplan–Meier method. Survival rates between patients treated with tocilizumab within two days after hospital admission and later, and between omicron and delta cases, were compared using the log-rank test. Then, survival rates between patients treated with tocilizumab + SoC (studied population) and SoC alone (control group), and between patients treated with tocilizumab within two days after hospital admission and later (studied population divided in two groups), and those treated with SoC alone (control group) were also compared using the log-rank test.

All tests were two-sided, and a P-value less than 0.05 was considered statistically significant. Analyses were performed using the R software version 4.2.0 (The R Foundation for Statistical Computing, Vienna, Austria).

5. Conclusions

In a cohort of severe COVID-19 patients infected with delta and omicron VOC and treated with tocilizumab in a French teaching hospital, we reported a 22% mortality. Mortality rates were similar in the omicron and delta cases, and in COVID-19 patients treated with tocilizumab plus SoC and SoC alone. However, our results suggested that the early administration of tocilizumab had a protective effect on mortality in severe COVID-19 patients regardless of the VOC involved. In contrast, a higher age, higher Charlson index and higher WHO-CPS were significantly associated with mortality. Further multicenter studies are awaited to confirm our results.

Author Contributions: Conceptualization, R.L.; methodology, P.L.-L. and R.L.; validation, A.S.; formal analysis, R.L.; investigation, P.L.-L., D.L., P.L., R.S., M.C., E.C., A.M., C.R. (Claire Roger), L.M., P.-G.C., R.G., C.R. (Clarisse Roux), J.-P.L. and R.L.; writing—original draft preparation, P.L.-L.; writing—review and editing, P.L., D.L., A.S. and R.L. All authors have read and agreed to the published version of the manuscript.

Funding: This research received no external funding.

Institutional Review Board Statement: The study was conducted in accordance with the Declaration of Helsinki and approved by the Institutional Review Board of Nimes University Hospital (No. 220115, date of approval: 2 October 2022).

Informed Consent Statement: Patient written consent was waived by the Institutional Review Board of Nimes University Hospital in accordance with the national legislation and institutional requirements because this observational study did not modify existing diagnostic or therapeutic strategies. However, patients were informed of their inclusion in the study.

Data Availability Statement: The authors consent to share the collected data with others. The raw data supporting the conclusions of this article will be made available by the authors, without undue reservation. Data will be available immediately after the main publication and indefinitely.

Conflicts of Interest: The authors declare no conflict of interest.

References

1. Moore, J.B.; June, C.H. Cytokine Release Syndrome in Severe COVID-19. *Science* **2020**, *368*, 473–474. [CrossRef] [PubMed]
2. Chen, X.; Zhao, B.; Qu, Y.; Chen, Y.; Xiong, J.; Feng, Y.; Men, D.; Huang, Q.; Liu, Y.; Yang, B.; et al. Detectable Serum Severe Acute Respiratory Syndrome Coronavirus 2 Viral Load (RNAemia) Is Closely Correlated With Drastically Elevated Interleukin 6 Level in Critically Ill Patients with Coronavirus Disease 2019. *Clin. Infect. Dis.* **2020**, *71*, 1937–1942. [CrossRef] [PubMed]
3. The RECOVERY Collaborative Group Dexamethasone in Hospitalized Patients with COVID-19. *N. Engl. J. Med.* **2021**, *384*, 693–704. [CrossRef]
4. Zhang, S.; Li, L.; Shen, A.; Chen, Y.; Qi, Z. Rational Use of Tocilizumab in the Treatment of Novel Coronavirus Pneumonia. *Clin. Drug. Investig.* **2020**, *40*, 511–518. [CrossRef] [PubMed]
5. Abani, O.; Abbas, A.; Abbas, F.; Abbas, M.; Abbasi, S.; Abbass, H.; Abbott, A.; Abdallah, N.; Abdelaziz, A.; Abdelfattah, M.; et al. Tocilizumab in Patients Admitted to Hospital with COVID-19 (RECOVERY): A Randomised, Controlled, Open-Label, Platform Trial. *Lancet* **2021**, *397*, 1637–1645. [CrossRef]
6. The REMAP-CAP Investigators Interleukin-6 Receptor Antagonists in Critically Ill Patients with COVID-19. *N. Engl. J. Med.* **2021**, *384*, 1491–1502. [CrossRef]
7. Cassone, G.; Dolci, G.; Besutti, G.; Braglia, L.; Pavone, P.; Corsini, R.; Sampaolesi, F.; Iotti, V.; Teopompi, E.; Massari, M.; et al. Predictive Factors of Clinical Outcomes in Patients with COVID-19 Treated with Tocilizumab: A Monocentric Retrospective Analysis. *PLoS ONE* **2022**, *17*, e0262908. [CrossRef]
8. Radulescu, A.; Istrate, A.; Muntean, M. Treatment with Tocilizumab in Adult Patients with Moderate to Critical COVID-19 Pneumonia: A Single-Center Retrospective Study. *Int. J. Infect. Dis.* **2022**, *117*, 1–7. [CrossRef]
9. Hafez, W.; Abdelrahman, A. Factors Influencing Disease Stability and Response to Tocilizumab Therapy in Severe COVID-19: A Retrospective Cohort Study. *Antibiotics* **2022**, *11*, 1078. [CrossRef]
10. San-Juan, R.; Fernández-Ruiz, M.; López-Medrano, F.; Carretero, O.; Lalueza, A.; Maestro de la Calle, G.; Pérez-Jacoiste Asín, M.A.; Bueno, H.; Caro-Teller, J.M.; Catalán, M.; et al. Analysis of the Factors Predicting Clinical Response to Tocilizumab Therapy in Patients with Severe COVID-19. *Int. J. Infect. Dis.* **2022**, *117*, 56–64. [CrossRef]
11. Nigo, M.; Rasmy, L.; May, S.B.; Rao, A.; Karimaghaei, S.; Kannadath, B.S.; De la Hoz, A.; Arias, C.A.; Li, L.; Zhi, D. Real World Long-Term Assessment of The Efficacy of Tocilizumab in Patients with COVID-19: Results from A Large De-Identified Multicenter Electronic Health Record Dataset in the United States. *Int. J. Infect. Dis.* **2021**, *113*, 148–154. [CrossRef] [PubMed]
12. Writing Committee for the REMAP-CAP Investigators; Florescu, S.; Stanciu, D.; Zaharia, M.; Kosa, A.; Codreanu, D.; Kidwai, A.; Masood, S.; Kaye, C.; Coutts, A.; et al. Long-Term (180-Day) Outcomes in Critically Ill Patients With COVID-19 in the REMAP-CAP Randomized Clinical Trial. *JAMA* **2022**. [CrossRef] [PubMed]
13. Rezaei Tolzali, M.M.; Noori, M.; Shokri, P.; Rahmani, S.; Khanzadeh, S.; Nejadghaderi, S.A.; Fazlollahi, A.; Sullman, M.J.M.; Singh, K.; Kolahi, A.-A.; et al. Efficacy of Tocilizumab in the Treatment of COVID-19: An Umbrella Review. *Rev. Med. Virol.* **2022**, *32*, e2388. [CrossRef] [PubMed]
14. Oliynyk, O.; Barg, W.; Oliynyk, Y.; Dubrov, S.; Gurianov, V.; Rorat, M. Lack of Difference in Tocilizumab Efficacy in the Treatment of Severe COVID-19 Caused by Different SARS-CoV-2 Variants. *J. Pers. Med.* **2022**, *12*, 1103. [CrossRef] [PubMed]

15. Ullah, S.; Abid, R.; Haider, S.; Khuda, F.; Albadrani, G.M.; Abdulhakim, J.A.; Altyar, A.E.; Abdel-Daim, M.M.; Halimi, S.M.A. Khalil, A.A.K. Assessment of Tocilizumab (Humanized Monoclonal Antibody) for Therapeutic Efficacy and Clinical Safety in Patients with Coronavirus Disease (COVID-19). *Medicina* **2022**, *58*, 1076. [CrossRef]
16. Wolter, N.; Jassat, W.; Walaza, S.; Welch, R.; Moultrie, H.; Groome, M.; Amoako, D.G.; Everatt, J.; Bhiman, J.N.; Scheepers, C.; et al. Early Assessment of the Clinical Severity of the SARS-CoV-2 Omicron Variant in South Africa: A Data Linkage Study. *Lancet* **2022** *399*, 437–446. [CrossRef]
17. Williamson, E.J.; Walker, A.J.; Bhaskaran, K.; Bacon, S.; Bates, C.; Morton, C.E.; Curtis, H.J.; Mehrkar, A.; Evans, D.; Inglesby, P. et al. Factors Associated with COVID-19-Related Death Using OpenSAFELY. *Nature* **2020**, *584*, 430–436. [CrossRef]
18. Lv, F.; Gao, X.; Huang, A.H.; Zu, J.; He, X.; Sun, X.; Liu, J.; Gao, N.; Jiao, Y.; Keane, M.G.; et al. Excess Diabetes Mellitus-Related Deaths during the COVID-19 Pandemic in the United States. *eClinicalMedicine* **2022**, *54*, 100387. [CrossRef]
19. Katz, A.; Altshuler, D.; Papadopoulos, J.; Amoroso, N.; Goldenberg, R.; Tarras, E.; Krolikowski, K.; Hagedorn, J.; Fridman, D. Chen, X.J.C.; et al. The Use of High-Dose Corticosteroids Versus Low-Dose Corticosteroids with and without Tocilizumab in COVID-19 Acute Respiratory Distress Syndrome. *Ann. Pharm.* **2023**, *57*, 5–15. [CrossRef]
20. Twohig, K.A.; Nyberg, T.; Zaidi, A.; Thelwall, S.; Sinnathamby, M.A.; Aliabadi, S.; Seaman, S.R.; Harris, R.J.; Hope, R.; Lopez-Bernal, J.; et al. Hospital Admission and Emergency Care Attendance Risk for SARS-CoV-2 Delta (B.1.617.2) Compared with Alpha (B.1.1.7) Variants of Concern: A Cohort Study. *Lancet Infect. Dis.* **2022**, *22*, 35–42. [CrossRef]
21. Nyberg, T.; Ferguson, N.M.; Nash, S.G.; Webster, H.H.; Flaxman, S.; Andrews, N.; Hinsley, W.; Bernal, J.L.; Kall, M.; Bhatt, S.; et al Comparative Analysis of the Risks of Hospitalisation and Death Associated with SARS-CoV-2 Omicron (B.1.1.529) and Delta (B.1.617.2) Variants in England: A Cohort Study. *Lancet* **2022**, *399*, 1303–1312. [CrossRef] [PubMed]
22. Korobova, Z.R.; Arsentieva, N.A.; Liubimova, N.E.; Batsunov, O.K.; Dedkov, V.G.; Gladkikh, A.S.; Sharova, A.A.; Adish, Z.; Chernykh, E.I.; Kaschenko, V.A.; et al. Cytokine Profiling in Different SARS-CoV-2 Genetic Variants. *Int. J. Mol. Sci.* **2022**, *23*, 14146. [CrossRef] [PubMed]
23. Richier, Q.; Jachiet, V.; Bonnemains, V.; Plaçais, L.; Abisror, N.; Garnier, M.; Pacanowski, J.; Dhote, R.; Hinchschberger, O.; Michel, M.; et al. Tocilizumab and COVID-19: Timing of Administration Assessment. *Infect. Dis. Now* **2022**, *52*, 31–34. [CrossRef]
24. Stone, J.H.; Frigault, M.J.; Serling-Boyd, N.J.; Fernandes, A.D.; Harvey, L.; Foulkes, A.S.; Horick, N.K.; Healy, B.C.; Shah, R.; Bensaci, A.M.; et al. Efficacy of Tocilizumab in Patients Hospitalized with COVID-19. *N. Engl. J. Med.* **2020**, *383*, 2333–2344. [CrossRef] [PubMed]
25. Rosas, I.O.; Bräu, N.; Waters, M.; Go, R.C.; Hunter, B.D.; Bhagani, S.; Skiest, D.; Aziz, M.S.; Cooper, N.; Douglas, I.S.; et al. Tocilizumab in Hospitalized Patients with Severe COVID-19 Pneumonia. *N. Engl. J. Med.* **2021**, *384*, 1503–1516. [CrossRef]
26. Salama, C.; Han, J.; Yau, L.; Reiss, W.G.; Kramer, B.; Neidhart, J.D.; Criner, G.J.; Kaplan-Lewis, E.; Baden, R.; Pandit, L.; et al. Tocilizumab in Patients Hospitalized with COVID-19 Pneumonia. *N. Engl. J. Med.* **2021**, *384*, 20–30. [CrossRef] [PubMed]
27. Sandhu, G.; Piraino, S.T.; Piticaru, J. Secondary Infection Risk in Patients with Severe COVID-19 Pneumonia Treated with Tocilizumab. *Am. J. Ther.* **2022**, *29*, e275–e278. [CrossRef]
28. Prattes, J.; Wauters, J.; Giacobbe, D.R.; Salmanton-García, J.; Maertens, J.; Bourgeois, M.; Reynders, M.; Rutsaert, L.; Van Regenmortel, N.; Lormans, P.; et al. Risk Factors and Outcome of Pulmonary Aspergillosis in Critically Ill Coronavirus Disease 2019 Patients—A Multinational Observational Study by the European Confederation of Medical Mycology. *Clin. Microbiol. Infect.* **2021**, *28*, 580–587. [CrossRef]
29. Ripa, M.; Galli, L.; Poli, A.; Oltolini, C.; Spagnuolo, V.; Mastrangelo, A.; Muccini, C.; Monti, G.; De Luca, G.; Landoni, G.; et al. Secondary Infections in Patients Hospitalized with COVID-19: Incidence and Predictive Factors. *Clin. Microbiol. Infect.* **2021**, *27*, 451–457. [CrossRef]
30. Langford, B.J.; So, M.; Raybardhan, S.; Leung, V.; Westwood, D.; MacFadden, D.R.; Soucy, J.-P.R.; Daneman, N. Bacterial Co-Infection and Secondary Infection in Patients with COVID-19: A Living Rapid Review and Meta-Analysis. *Clin. Microbiol. Infect.* **2020**, *26*, 1622–1629. [CrossRef]
31. Koehler, P.; Bassetti, M.; Chakrabarti, A.; Chen, S.C.A.; Colombo, A.L.; Hoenigl, M.; Klimko, N.; Lass-Flörl, C.; Oladele, R.O.; Vinh, D.C.; et al. Defining and Managing COVID-19-Associated Pulmonary Aspergillosis: The 2020 ECMM/ISHAM Consensus Criteria for Research and Clinical Guidance. *Lancet Infect. Dis.* **2021**, *21*, e149–e162. [CrossRef] [PubMed]
32. Khurshid, S.; Rehman, N.; Ahmed, S.; Ahmad, B.; Khurshid, M.; Muhammad, A.; Siddiqi, F.A.; Nayab, D.; Saleem, H.; Saleem, Z. Early Fall in C-Reactive Protein (CRP) Level Predicts Response to Tocilizumab in Rapidly Progressing COVID-19: Experience in a Single-Arm Pakistani Center. *Cureus* **2021**, *13*, e20031. [CrossRef] [PubMed]
33. *Therapeutics and COVID-19: Living Guideline, 22 April 2022*; World Health Organization: Geneva, Switzerland, 2022.
34. Albuquerque, A.M.; Tramujas, L.; Sewanan, L.R.; Williams, D.R.; Brophy, J.M. Mortality Rates among Hospitalized Patients with COVID-19 Infection Treated with Tocilizumab and Corticosteroids: A Bayesian Reanalysis of a Previous Meta-Analysis. *JAMA Netw. Open* **2022**, *5*, e220548. [CrossRef] [PubMed]
35. The WHO Rapid Evidence Appraisal for COVID-19 Therapies (REACT) Working Group Association between Administration of IL-6 Antagonists and Mortality among Patients Hospitalized for COVID-19: A Meta-Analysis. *JAMA* **2021**, *326*, 499–518. [CrossRef]
36. COVID-19 Treatment Guidelines Panel Coronavirus Disease 2019 (COVID-19) Treatment Guidelines. Available online: https://www.covid19treatmentguidelines.nih.gov (accessed on 23 November 2022).

37. European Centre for Disease Prevention and Control Strategies for the Surveillance of COVID-19 2020. Available online: https://www.ecdc.europa.eu/en/publications-data/strategies-surveillance-covid-19 (accessed on 23 November 2022).
38. Ojha, V.; Mani, A.; Pandey, N.N.; Sharma, S.; Kumar, S. CT in Coronavirus Disease 2019 (COVID-19): A Systematic Review of Chest CT Findings in 4410 Adult Patients. *Eur. Radiol.* **2020**, *30*, 6129–6138. [CrossRef]
39. International Classification of Diseases (ICD). Available online: https://www.who.int/standards/classifications/classification-of-diseases (accessed on 6 June 2022).
40. Charlson, M.E.; Pompei, P.; Ales, K.L.; MacKenzie, C.R. A New Method of Classifying Prognostic Comorbidity in Longitudinal Studies: Development and Validation. *J. Chronic. Dis.* **1987**, *40*, 373–383. [CrossRef] [PubMed]
41. Abdel-Tawab, M.; Basha, M.A.A.; Mohamed, I.A.I.; Ibrahim, H.M. A Simple Chest CT Score for Assessing the Severity of Pulmonary Involvement in COVID-19. *Egypt. J. Radiol. Nucl. Med.* **2021**, *52*, 149. [CrossRef]
42. Marshall, J.C.; Murthy, S.; Diaz, J.; Adhikari, N.K.; Angus, D.C.; Arabi, Y.M.; Baillie, K.; Bauer, M.; Berry, S.; Blackwood, B.; et al. A Minimal Common Outcome Measure Set for COVID-19 Clinical Research. *Lancet Infect. Dis.* **2020**, *20*, e192–e197. [CrossRef]

antibiotics MDPI

Systematic Review

Increasing Consumption of Antibiotics during the COVID-19 Pandemic: Implications for Patient Health and Emerging Anti-Microbial Resistance

Shahana Seher Malik [1] and Sunil Mundra [1,2,*]

1 Department of Biology, College of Science, United Arab Emirates University, Al Ain P.O. Box 15551, United Arab Emirates
2 Khalifa Center for Genetic Engineering and Biotechnology, United Arab Emirates University, Al Ain P.O. Box 15551, United Arab Emirates
* Correspondence: sunilmundra@uaeu.ac.ae; Tel.: +971-7136341

Abstract: The emergence of COVID-19 infection led to the indiscriminate use of antimicrobials without knowing their efficacy in treating the disease. The gratuitous use of antibiotics for COVID-19 treatment raises concerns about the emergence of antimicrobial resistance (AMR). In this systematic review, we performed a thorough systematic search using Preferred Reporting Items for Systematic Reviews and Meta-Analyses (PRISMA) guidelines of scientific databases (Scopus, Web of Science, and PubMed) to identify studies where antibiotics were prescribed to treat COVID-19 (December 2019 to December 2021). Of 970 identified studies, 130 were included in our analyses. Almost 78% of COVID-19 patients have been prescribed an antibiotic. Cephalosporins were the most prescribed (30.1% of patients) antibiotics, followed by azithromycin (26% of patients). Antibiotics were prescribed for COVID-19 patients regardless of reported severity; the overall rate of antibiotic use was similar when comparing patients with a severe or critical illness (77.4%) and patients with mild or moderate illness (76.8%). Secondary infections were mentioned in only 11 studies. We conclude that concerns related to COVID-19 and the lack of treatment strategy led to the overuse of antibiotics without proper clinical rationale. Based on our findings, we propose that antimicrobial stewardship should be retained as a priority while treating viral pandemics.

Keywords: AMR; antibiotics; cephalosporin; COVID-19; co-infection; secondary infection

Citation: Malik, S.S.; Mundra, S. Increasing Consumption of Antibiotics during the COVID-19 Pandemic: Implications for Patient Health and Emerging Anti-Microbial Resistance. *Antibiotics* **2023**, *12*, 45. https://doi.org/10.3390/antibiotics12010045

Academic Editors: Sabine Danielle Allard and Johan Van Laethem

Received: 23 November 2022
Revised: 24 December 2022
Accepted: 26 December 2022
Published: 28 December 2022

1. Introduction

During the current coronavirus disease 2019 (COVID-19) pandemic, antimicrobial use increased tremendously due to the lack of proper treatment strategies [1]. Although COVID-19 is a viral disease that is untreatable by antibiotics, viral respiratory infections can progress to bacterial pneumonia, co-infection, bacterial superinfection, and other nosocomial infections requiring antibiotic administration.

The overuse of antibiotics for treating COVID-19 patients was a result of (a) panic about an unknown disease, (b) similar symptoms to pneumonia, and (c) a higher death rate in communities with weaker immunity [2–4]. Moreover, the bacterial co-infection rate was almost 16%, and the use of antibiotics (especially broad-spectrum) increased to more than 72% during the current pandemic [3].

The consumption of antibiotics during COVID-19 increased tremendously, and there are various factors contributing to the spread of AMR [5,6]. Enhanced hospital exposure is silently contributing to the emerging rate of antimicrobial resistance (AMR), which causes about 700,000 deaths per year globally [7]. The widespread use of antibiotics to control pandemics might also increase resistant organisms [8,9]. The current state of antibiotic use in COVID-19 projects approximately 10 million deaths by 2050 [9–13]. The evidence suggests that the COVID-19 pandemic is increasing the rate of AMR through the unnecessary use of antibiotics [13]. Therefore, it is critical to strengthen antimicrobial stewardship (AMS) and formulate policies for the use of antibiotics [12].

The driving factors for the use of antibiotics include lack of proper awareness in the public, potential access and affordability to antibiotics without prescription, and use of leftover antibiotics from earlier prescriptions. While some other driving factors are insufficient training during the early phase of clinical practice, irrational prescriptions to promote a pharmaceutical company, and inadequate diagnostic process [14].

The lack of new antibiotics development for the past three decades, termed as "discovery void", is due to the scarcity of research to find new antimicrobials [12,14]. COVID-19 has also disrupted the production, delivery, and processing of antimicrobials. During the current pandemic, the demand to find a treatment has led to a shift in research resources and funds to new antivirals and vaccines instead of antibiotics [15]. Furthermore, the shortage and rerouting of medical funds during COVID-19 have affected many small labs that produce medicines and vaccines for local markets. This deficiency of narrow-spectrum antibiotics can cause an increase in AMR [16]. Clinical trials were also disrupted as the hospitals focused on COVID-19 [17]. Further, there is a need to start clinical trials of many antimicrobials. The results of these trials are expected to improve COVID-19 treatments and patient outcomes.

The studies focusing on AMR have raised concern about the inflammatory effects of administered drugs [18]. Therefore, different countries have formulated many guidelines for antimicrobial use during the pandemic. However, the World Health Organization (WHO) has recommended avoiding antibiotic use for patients with mild to moderate symptoms of bacterial or COVID-19 infections [19]. According to WHO, for severe cases, only low-potency antibiotics are recommended, and in the case of aged persons, the antibiotics included in the access list of WHO (https://aware.essentialmeds.org/groups (accessed on 25 August 2021)) can only be prescribed [20]. Nonetheless, these guidelines are insufficient to limit AMR emergence. There is a strict need for evidence-based guidelines for AMS during and/or post-COVID-19 pandemic. There is a need to analyze the trends for the pandemic's spread and the complete details of antibiotics used globally since the COVID-19 outbreak.

Here, we performed a systematic analysis to assess changes in antibiotic use during the COVID-19 era and how these changes might impact AMR.

2. Results

Based on our keyword search using PRISMA, we selected 970 research articles. Of these, 260 articles (84 review studies, 172 journal articles, and 4 short surveys) were further screened based on the exclusion and inclusion criteria. We further excluded another 130 studies that did not report the application of antimicrobials while treating COVID-19 patients. Finally, a total of 130 articles were selected for final systematic synthesis (Figure 1, Table S1). Among selected studies, most were conducted in the USA, followed by the UK, India, Italy, and China.

2.1. *Usage of Antibiotics according to the Severity of Illness*

Overall, 47.6% of patients were suffering from severe or critical illness, while the remaining cases were of mild or moderate nature. Almost 78% of patients were prescribed antibiotics. A minor difference was seen in the prescription of antibiotics among severe or critical and moderate or mild patients (77.4% and 76.8%; Figure 2).

2.2. *Use of Antibiotics and Related Health Effects*

The mortality rate was higher in cases where all patients used antibiotics than in cases in which most patients were not given antibiotics. Length of hospital stay (LOS) was higher in the patients' group, where not all, but the majority, were given antibiotics. The discharge rate was highest amongst those patients who were not given antibiotics compared to the group where most of the patients received antibiotics (Figure 3).

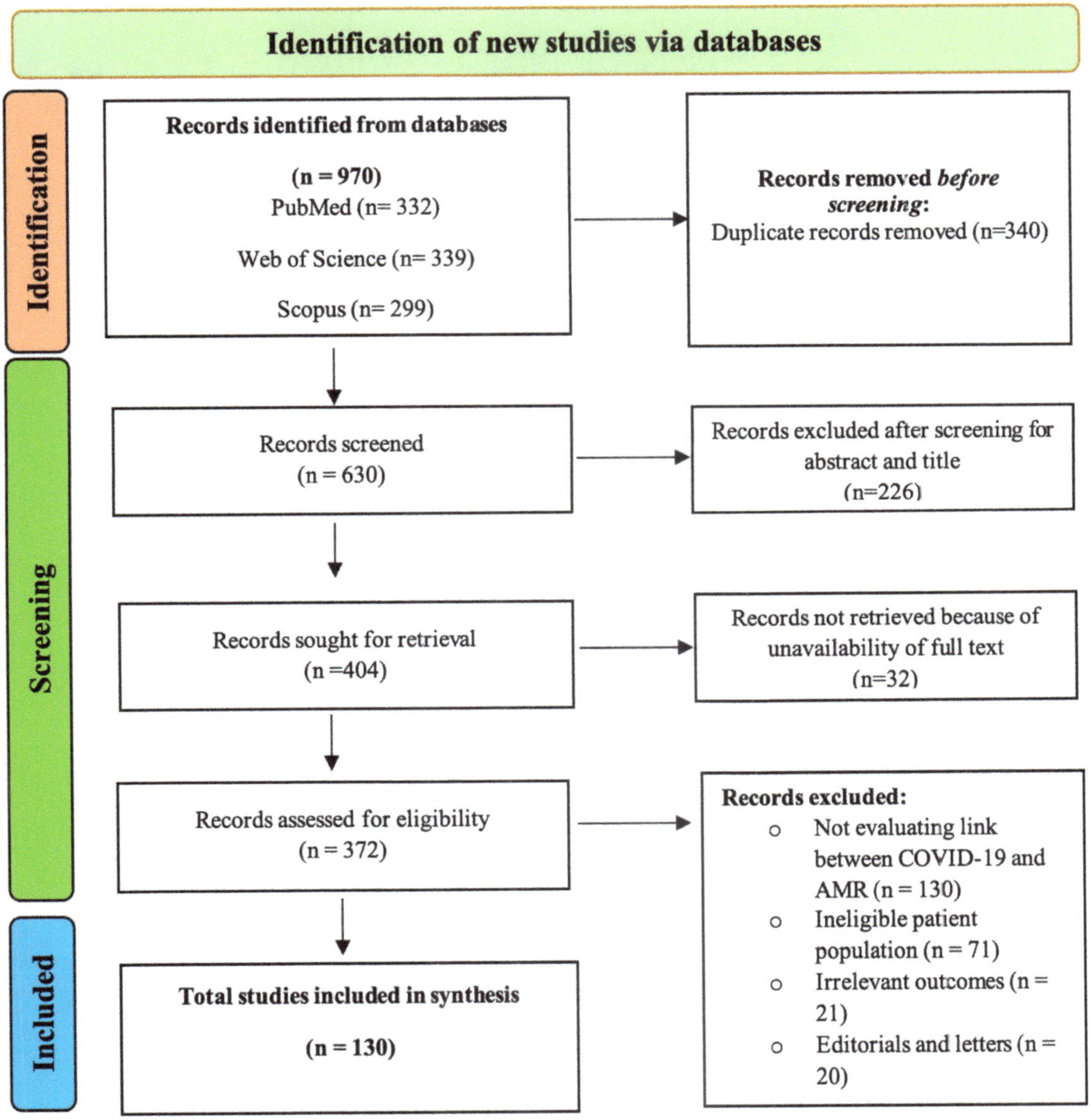

Figure 1. PRISMA workflow for the literature identification from 3 databases (PubMed, Web of Science, and Scopus), screening, and final selection of research studies for synthesis study. The exclusion criteria of the workflow include duplicate studies from multiple databases, screening based on the irrelevance of title and abstract, removal of records due to unavailability of full-text studies, studies not evaluating the link between COVID-19 and AMR, and having irrelevant outcomes, editorials, and letters.

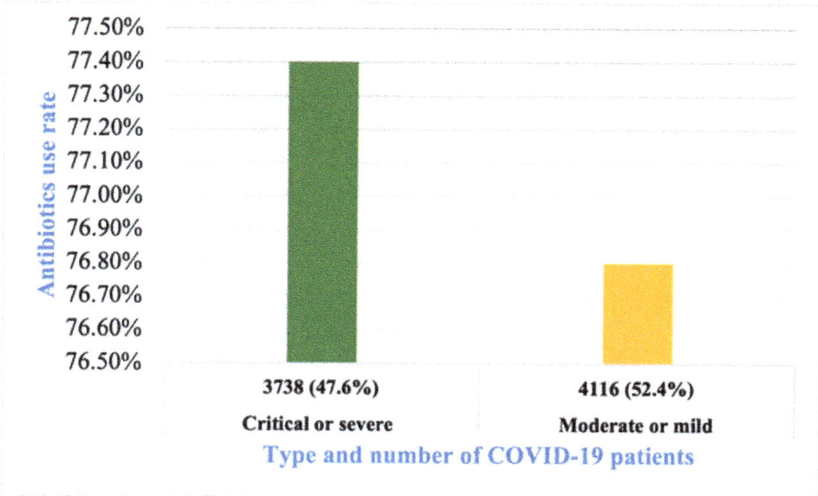

Figure 2. Rate of antibiotic use in COVID-19 patients according to the severity of illness.

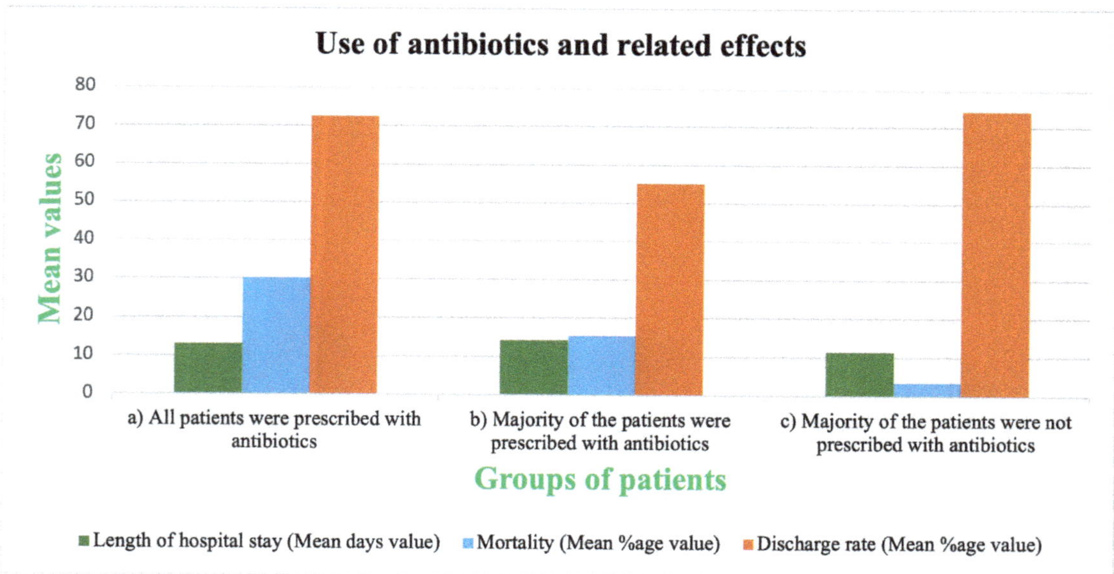

Figure 3. The use of antibiotics in COVID-19 patients divided into three groups (**a**) all patients using antibiotics, (**b**) the majority of the patients using antibiotics, and (**c**) the majority of the patients not using antibiotics) and related effects (length of hospital stay, mortality, and discharge rate).

2.3. Bacterial Co-Infections and Resulting Health Outputs

Of the total, 11 studies reported secondary infections in COVID-19 patients. In our analysis, 51.3% of the patients showed secondary infections, and 57.5% were in critical condition (Table 1).

Table 1. The comparison of disease severity, mean length of stay in hospital, mean discharge rate, and mean mortality rate among patients with secondary infections and normal COVID-19 patients.

Serial No.	Category	Critical or Severe n (%age)	Moderate or Mild *n* (%)	Mean Length of Stay (Days)	Mean Discharge Rate (%age)	Mean Mortality Rate (%age)
1	Patients with secondary infections (*n* = 770)	443 (57.5%)	327 (42.5%)	20.8	53.9	46.5
2	Total patients (*n* = 6512)	2807 (43.2%)	3705 (56.8%)	12.7	61.7	18.3

Among all patients, the percentage of confirmed co-infections was 12%. The patients with secondary infections also had a higher mean value of LOS and mortality rate. At the same time, the hospital discharge rate was lower in these patients (Figure 4).

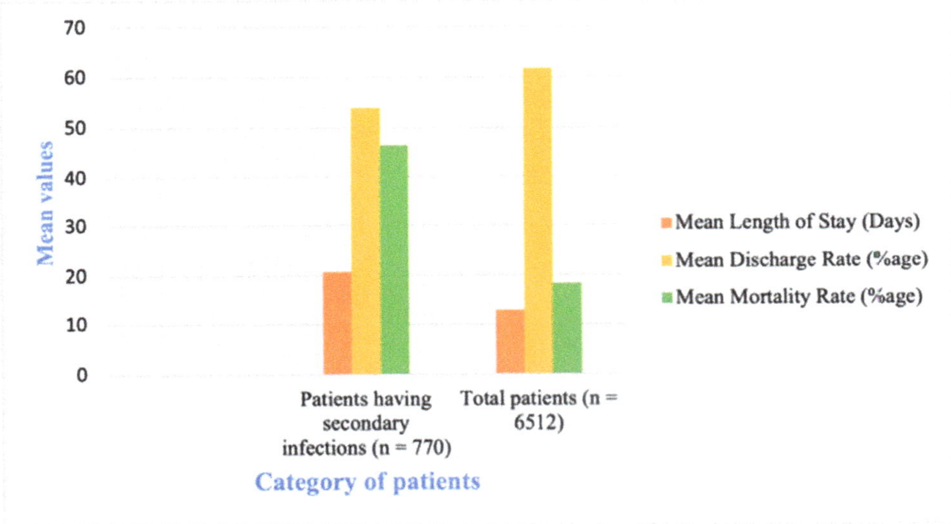

Figure 4. Comparison of hospital stay, discharge rate, and mortality rate in COVID-19 patients with and without secondary infections.

2.4. Antimicrobials' Usage Trends

The number of antimicrobial types prescribed showed a very slight increase in 2020 and 2021 compared to 2019, but the number of patients using antimicrobials increased tremendously over that time. In 2019 the number of different antimicrobials being used was 645, and in 2021 it increased by almost 4-fold to 2503. The total number of admitted patients and available beds in the surveyed wards also increased simultaneously from the year 2019 to the year 2021 as the number of COVID-19 cases surged (Figure 5).

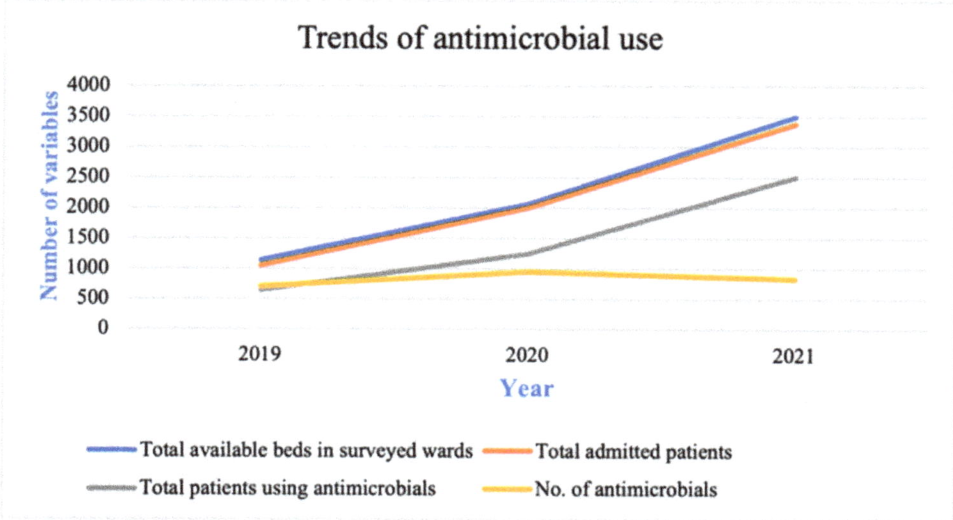

Figure 5. Trends of antimicrobial use in hospitalized COVID-19 patients along with total available beds in surveyed hospital wards.

3. Discussion

The extensive use of antibiotics has increased since the COVID-19 outbreak, increasing concerns about AMR. This study includes only the data of hospitalized patients because very little data are available about those confined to their homes with mild or moderate symptoms. This is probably because most research studies are performed on patients reporting to medical facilities.

We found that >78% of COVID-19 patients were recommended to use antibiotics (Figure 2). The broad-spectrum antibiotic azithromycin was the most frequently prescribed, followed by ceftriaxone, moxifloxacin, meropenem, and tazobactam. The excessive use of broad-spectrum antibiotics without proper clinical justifications by healthcare persons has raised concerns about AMR amplification [21].

It has been previously reported that azithromycin was the most common antimicrobial agent used while treating COVID-19 [22]. Azithromycin is very efficient in treating pneumonia, but there is no proof of its effect on viruses. There are studies that clearly show that excessive use of azithromycin can result in antimicrobial resistance, which indicates that overuse or misuse of this antibiotic during COVID-19 can contribute to AMR [23,24]. The WHO also reported the excess use of azithromycin for COVID-19 treatment even without approval [25]. It is classified as a critically important antimicrobial used in humans [26]. Persistent use of this antimicrobial can result in AMR, causing a serious threat to survival in severe infection cases. Third-generation cephalosporins such as ceftriaxone have long been mostly used in intensive care units (ICUs). As the consumption of other antibiotics changed during the pandemic, cephalosporins were also being repurposed [27]. Karami et al. provided clear evidence of cephalosporin use in more than 200 of 556 cases [28].

Doxycycline is another drug that could help treat COVID-19, which is used because of its antiviral and anti-inflammatory properties [29]. Many broad-spectrum antibiotics, such as fluoroquinolones and cephalosporins, were reportedly used in more than 74% of patients. If this use of broad-spectrum antibiotics is not stopped, it could lead to AMR. This could lead to very few remaining options for treating infections, and these options could become unaffordable for underdeveloped countries.

Increasing telehealth services also caused excessive antibiotic use due to the unavailability of the proper diagnostic channel and determining the nature of the illness [30]. The best treatment options for COVID-19 were unknown, so experimental treatments played a large role in inappropriate drug prescriptions during this pandemic period [17]. Due to this vagueness in the COVID-19 era, healthcare workers prescribed antibiotics based on the assumption that the drug's potential threat would be negligible compared to its benefit [31]. Previous studies also mentioned increased antimicrobial use due to self-medication [32–34]. This was more common in areas where antimicrobials could be easily accessed without any prescription [16,35]. One study reported that to avoid visits to any healthcare facility, about 20% of Iranians used self-prescribed medicines for their sickness [36].

We found that the recommendation rate of antibiotics does not change with the severity of the disease (Figure 2). Both mild or moderate and severe or critical illness patient groups were given antibiotics, although the severe cases had a greater chance of suffering from a secondary infection. The high rate of antibiotic application in mild cases is also very alarming. The results show that the mortality rate was higher in cases where all patients used antibiotics than in cases in which most patients were not given antibiotics. Length of hospital stay (LOS) was higher in the patients' group, where not all, but the majority, were given antibiotics (Figure 3). In our analysis, 51.3% of the patients showed secondary infections, and 57.5% were in critical condition (Table 1). We also found that the co-infection rate was higher in patients with severe COVID-19 symptoms, and the mortality rate was greater in patients with some co-infection or secondary infection (Figure 4). Although the rate of antibiotic prescription is enormous, secondary infections were only reported in 14.3% of cases, along with 3.5% of reported cases of co-infections [20]. The bacterial co-infection of COVID-19 patients has been reported in many studies worldwide [37–41]. The co-infection rate was almost 28% in Europe [42]. Patients with mild or medium symptoms were not reported for co-infection because these patients were not checked or tested for infection [43–45]. Moreover, in many cases, when the specimen was taken, the co-infections were reported to be more associated with hospital-acquired than community-acquired infections.

Our analysis shows that the number of antimicrobial types increased only slightly in 2020 and 2021 compared to 2019, but the number of patients using antimicrobials increased tremendously (a four-fold increase) (Figure 5). The total number of admitted patients also increased from 2019 to 2021. This indicates a positive relationship between the number of COVID-19 patients and the antimicrobials consumption, contributing to AMR. A higher rate of AMR could be predicted in low- and middle-income countries because of a lack of awareness and stewardship programs, poor lab facilities, and a lack of proper rules for accessing antibiotics without prescription [46]. COVID-19 can be more easily spread to areas that are more populated and lack proper hygiene facilities. In low- and middle-income countries, allocating resources to COVID-19 is very difficult as their healthcare facilities already lack proper funds, which is an additional burden to their healthcare systems [32].

In many studies, classes of antibiotics are mentioned, but information related to the use of antibiotics according to disease severity was not available. The variation in the number of research studies across regions might have affected our results, as there is a huge difference in the number of patients and local regulations for COVID-19 [47]. Moreover, the selection biases (not all the studies included in this systematic review are directly discussing AMR and COVID-19) are there. In this paper, we mostly focused on selecting the papers that provide evidence of antibiotic use during COVID-19, as the data related to antibiotics use can help in providing a clear understanding of the rate of AMR prevalence at this time. Healthcare professionals were also under pressure during the COVID-19 pandemic. In small hospitals where healthcare facilities are insufficient, they must try every possible option to save the life of patients. The diagnosis of secondary infections is very costly, and most hospitals lack this facility, leading doctors to prescribe antibiotics even when they are not needed. This needs to be managed as the excessive use of antimicrobials raises the threat of AMR.

Antimicrobial Stewardship (AMS)

Our analysis demonstrates that COVID-19 has crucial implications for AMR. The overuse or misuse of antibiotics to control COVID-19 symptoms, even without any co-infection, has worsened the situation. The world health organization has clear guidelines in this regard; patients with mild symptoms should not be prescribed antibiotics. The antibiotics should only be prescribed when there is clear evidence of bacterial co-infection. Better diagnostics are required to identify patients with secondary infections to avoid misuse of antimicrobials. AMS programs can assist in properly using antimicrobials by reviewing every prescribed medication. The data to date show that the use of antimicrobials is much higher than needed. For real-time data review and dissemination, additional efforts are needed to improve AMS. These efforts will not only help control COVID-19 but will have a major role in controlling the future pandemic of AMR.

4. Materials and Methods

The study was conducted to identify and analyze the research studies reporting the use of antimicrobials (especially azithromycin, doxycycline, clarithromycin, ceftriaxone, erythromycin, amoxicillin, amoxicillin-clavulanic acid, ampicillin, gentamicin, benzylpenicillin) for treating COVID-19. The data were reported using Preferred Reporting Items for Systematic Reviews and Meta-Analyses (PRISMA) guidelines for systematic review, and its protocol is registered in the PROSPERO register [48,49]. In order to maximize the authenticity of our findings, we tried to include as many research studies as possible in this analysis. All data were from hospitalized COVID-19 patients, and the final selection was based on RT-PCR-confirmed COVID-19 cases.

4.1. Strategy for Data Search

The study included all papers from several scientific databases such as Scopus, Web of Science, and PubMed-associated peer-reviewed journals published since the COVID outbreak (December 2019 till December 2021). The keywords and search terms used were "Antimicrobial resistance" and "coronavirus" or "COVID-19" and "Antimicrobial resistance" and "Antimicrobial stewardship" or "Antibiotic resistance" and "COVID-19".

4.2. Enclosure and Elimination Criteria for Research Studies

We included all the studies: (1) related to COVID-19 patients from communities and hospitals; (2) where antibiotics were prescribed during treatment; and (3) published in English (or in press). We selected cohort studies, cross-sectional studies, case-control studies, randomized control studies, and descriptive and observational research studies related to the use of antimicrobials in COVID-19 patients. We included those studies reporting the patients and use of antibiotics without any discrimination in gender, age, color, country, or community. We also considered the studies that reported antibiotics usage without specifying their types or treatment outcomes. We excluded those research studies that overlapped (duplicate data), contained unreliable data (short reports containing no proper results), were published in the form of editorials and notes, and studies related to engineering and earth sciences. Our analysis did not consider the studies related to animal experimentation, molecular mechanism, drug modeling, and other aspects of COVID-19, except for the use of antibiotics/antimicrobials.

4.3. Data Extraction

After cross-checking for the study accuracy and duplicates, data extraction was carried out based on the year of study, type of article, design or idea of the study, area/country of study, the sternness of COVID-19, rate of bacterial co-infections (infection acquired with first 24 to 48 h of hospitalization) and secondary infections (infection acquired after 24 to 48 h of hospitalization), prescribed antimicrobials, and the number of patients consuming those antimicrobials. The extraction of data from selected research studies comprised of details of publication, region of study, number of reported patients, type of study (e.g., case study, cohort study, descriptive study), condition of patients (mild, severe, or critical), rate of antimicrobial usage and recommendation, the scenarios of prescription, time of hospital stay, and mortality rate. More than half of the studies reported disease severity as mild, moderate, critical, and severe, while the others used just mild, moderate, and severe. We categorized disease severity into two major groups: mild or moderate and severe or critical.

4.4. Data Synthesis and Analysis

We analyzed all the selected studies where details about medicine prescriptions were available. We also investigated antimicrobials and antibiotics use according to the severity of illness, types that were used most often to treat COVID-19, health effects related to the use of antibiotics, bacterial co-infections, and related outputs, and trends of antimicrobial usage at the time of COVID-19. We also explored the length of hospital stay (LOS), rate of discharge, and mortality (patients still in the hospital at the time of publication were excluded from this calculation).

5. Conclusions

The study reveals that most antibiotics (mainly azithromycin and cephalosporins) prescribed during COVID-19 treatment were not related to disease severity since antibiotics were prescribed to patients with mild symptoms. Moreover, antibiotics were recommended without any medical or biological indication of secondary bacterial infection and thus may not have been beneficial in remedying COVID-19 infection. The extensive use of antibiotics might augment antibiotic resistance, thereby eroding the usefulness of currently available antibiotics. The overuse of antibiotics may silently trigger future pandemics due to the emerging AMR [50].

Supplementary Materials: The following supporting information can be downloaded at: https://www.mdpi.com/article/10.3390/antibiotics12010045/s1, Table S1: List of studies included for final Systematic Review.

Author Contributions: Conceptualization, S.S.M. and S.M.; methodology, S.S.M.; formal analysis, S.S.M.; resources, S.M.; data curation, S.S.M.; writing—original draft preparation, S.S.M.; writing—review and editing, S.M.; visualization, S.S.M.; supervision, S.M.; project administration, S.M.; funding acquisition, S.M. All authors have read and agreed to the published version of the manuscript.

Funding: This work was financially supported by an ASPIRE grant (G00003604; #21S111; #AYIA20-003) from the United Arab Emirates University, Al Ain, United Arab Emirates.

Institutional Review Board Statement: Not applicable.

Informed Consent Statement: Not applicable.

Data Availability Statement: Not applicable.

Acknowledgments: The authors would like to thank the Abu Dhabi Young Investigator Award (AYIA) for facilitating our research idea.

Conflicts of Interest: The authors declare no conflict of interest.

References

1. Baby, B.; Devan, A.R.; Nair, B.; Nath, L.R. The Impetus of COVID-19 in Multiple Organ Affliction Apart from Respiratory Infection: Pathogenesis, Diagnostic Measures and Current Treatment Strategy. *Infect. Disord.-Drug Targets Former. Curr. Drug Targets-Infect. Disord.* **2021**, *21*, 514–526. [CrossRef]
2. Stankovska, G.; Memedi, I.; Dimitrovski, D. Coronavirus COVID-19 disease, mental health and psychosocial support. *Soc. Regist.* **2020**, *4*, 33–48. [CrossRef]
3. Townsend, L.; Hughes, G.; Kerr, C.; Kelly, M.; O'Connor, R.; Sweeney, E.; Doyle, C.; O'Riordan, R.; Martin-Loeches, I.; Bergin, C.; et al. Bacterial pneumonia coinfection and antimicrobial therapy duration in SARS-CoV-2 (COVID-19) infection. *JAC-Antimicrobial Resist.* **2020**, *2*, dlaa071. [CrossRef]
4. Wang, L.; Alexander, C.A. COVID-19 Compared with Other Viral Diseases: Novelties, Progress, and Challenges. *Electron. J. Gen. Med.* **2020**, *18*, em265. [CrossRef] [PubMed]

5. Mazumder, P.; Kalamdhad, A.; Chaminda, G.T.; Kumar, M. Coalescence of co-infection and antimicrobial resistance with SARS-CoV-2 infection: The blues of post-COVID-19 world. Case Stud. *Chem. Environ. Eng.* **2021**, *3*, 100093. [CrossRef]
6. Rasul, C. Indiscriminate use of antimicrobials during COVID-19 pandemic. *Bangladesh Med. J. Khulna* **2021**, *53*, 1–2. [CrossRef]
7. Machowska, A.; Lundborg, C.S. Drivers of Irrational Use of Antibiotics in Europe. *Int. J. Environ. Res. Public Health* **2018**, *16*, 27. [CrossRef] [PubMed]
8. Caselli, E. Hygiene: Microbial strategies to reduce pathogens and drug resistance in clinical settings. *Microb. Biotechnol.* **2017**, *10*, 1079–1083. [CrossRef]
9. Hashmi, F.K.; Atif, N.; Malik, U.R.; Saleem, F.; Riboua, Z.; Hassali, M.A.; Butt, M.H.; Mallhi, T.H.; Khan, Y.H. In Pursuit of COVID-19 Treatment Strategies: Are We Triggering Antimicrobial Resistance? *Disaster Med. Public Health Prep.* **2020**, *16*, 1285–1286. [CrossRef] [PubMed]
10. Cong, W.; Poudel, A.; Alhusein, N.; Wang, H.; Yao, G.; Lambert, H. Antimicrobial Use in COVID-19 Patients in the First Phase of the SARS-CoV-2 Pandemic: A Scoping Review. *Antibiotics* **2021**, *10*, 745. [CrossRef]
11. Baggs, J.; Rose, A.N.; McCarthy, N.L.; Wolford, H.; Srinivasan, A.; A Jernigan, J.; Reddy, S.C. Antibiotic-Resistant Infections Among Inpatients with Coronavirus Disease 2019 (COVID-19) in US Hospitals. *Clin. Infect. Dis.* **2022**, *75*, S294–S297. [CrossRef] [PubMed]
12. Getahun, H.; Smith, I.; Trivedi, K.; Paulin, S.; Balkhy, H.H. Tackling antimicrobial resistance in the COVID-19 pandemic. *Bull World Health Organ.* **2020**, *98*, 442–442A. [CrossRef] [PubMed]
13. A Strathdee, S.; Davies, S.C.; Marcelin, J.R. Confronting antimicrobial resistance beyond the COVID-19 pandemic and the 2020 US election. *Lancet* **2020**, *396*, 1050–1053. [CrossRef]
14. Zawahir, S.; Le, H.; Nguyen, T.A.; Beardsley, J.; Duc, A.D.; Bernays, S.; Viney, K.; Hung, T.C.; McKinn, S.; Tran, H.H.; et al. Standardised patient study to assess tuberculosis case detection within the private pharmacy sector in Vietnam. *BMJ Glob. Health* **2021**, *6*, e006475. [CrossRef] [PubMed]
15. Ukuhor, H.O. The interrelationships between antimicrobial resistance, COVID-19, past, and future pandemics. *J. Infect. Public Health* **2020**, *14*, 53–60. [CrossRef] [PubMed]
16. Khor, W.P.; Olaoye, O.; D'Arcy, N.; Krockow, E.M.; Elshenawy, R.A.; Rutter, V.; Ashiru-Oredope, D. The Need for Ongoing Antimicrobial Stewardship during the COVID-19 Pandemic and Actionable Recommendations. *Antibiotics* **2020**, *9*, 904. [CrossRef]
17. Nieuwlaat, R.; Mbuagbaw, L.; Mertz, D.; Burrows, L.L.; E Bowdish, D.M.; Moja, L.; Wright, G.D.; Schünemann, H.J. Coronavirus Disease 2019 and Antimicrobial Resistance: Parallel and Interacting Health Emergencies. *Clin. Infect. Dis.* **2020**, *72*, 1657–1659. [CrossRef]
18. Buetti, N.; Mazzuchelli, T.; Priore, E.L.; Balmelli, C.; Llamas, M.; Pallanza, M.; Elzi, L.; Consonni, V.; Trimboli, P.; Forni-Ogna, V.; et al. Early administered antibiotics do not impact mortality in critically ill patients with COVID-19. *J. Infect.* **2020**, *81*, e148–e149. [CrossRef]
19. World Health Organization. *Living Guidance for Clinical Management of COVID-19: Living Guidance*; World Health Organization: Geneva, Switzerland, 2021.
20. WHO. *Responding Community Spread COVID-19 Ref WHO COVID-19 Community Transmission 2020–2021*; World Health Organization: Geneva, Switzerland, 2020.
21. Tenforde, M.W.; Kim, S.S.; Lindsell, C.J.; Rose, E.B.; Shapiro, N.I.; Files, D.C.; Gibbs, K.W.; Erickson, H.L.; Steingrub, J.S.; Smithline, H.A.; et al. Symptom Duration and Risk Factors for Delayed Return to Usual Health among Outpatients with COVID-19 in a Multistate Health Care Systems Network—United States, March–June 2020. MMWR. *Morb. Mortal. Wkly. Rep.* **2020**, *69*, 993–998. [CrossRef]
22. Cangini, A.; Fortinguerra, F.; Di Filippo, A.; Pierantozzi, A.; Da Cas, R.; Villa, F.; Trotta, F.; Moro, M.L.; Gagliotti, C. Monitoring the community use of antibiotics in Italy within the National Action Plan on antimicrobial resistance. *Br. J. Clin. Pharmacol.* **2020**, *87*, 1033–1042. [CrossRef]
23. Hooda, Y.; Tanmoy, A.M.; Sajib, M.S.I.; Saha, S. Mass azithromycin administration: Considerations in an increasingly resistant world. *BMJ Glob. Health* **2020**, *5*, e002446. [CrossRef] [PubMed]
24. Mack, I.; Sharland, M.; A Berkley, J.; Klein, N.; Malhotra-Kumar, S.; Bielicki, J. Antimicrobial Resistance Following Azithromycin Mass Drug Administration: Potential Surveillance Strategies to Assess Public Health Impact. *Clin. Infect. Dis.* **2019**, *70*, 1501–1508. [CrossRef]
25. E Lane, J.C.; Weaver, J.; Kostka, K.; Duarte-Salles, T.; Abrahao, M.T.F.; Alghoul, H.; Alser, O.; Alshammari, T.M.; Biedermann, P.; Banda, J.M.; et al. Risk of hydroxychloroquine alone and in combination with azithromycin in the treatment of rheumatoid arthritis: A multinational, retrospective study. *Lancet Rheumatol.* **2020**, *2*, e698–e711. [CrossRef] [PubMed]
26. World Health Organization. *Critically Important Antimicrobials for Human Medicine*, 6th ed.; World Health Organization: Geneva, Switzerland, 2019. Available online: https://apps.who.int/iris/handle/10665/312266 (accessed on 25 August 2021).
27. Durojaiye, A.B.; Clarke, J.-R.D.; Stamatiades, G.A.; Wang, C. Repurposing cefuroxime for treatment of COVID-19: A scoping review of in silico studies. *J. Biomol. Struct. Dyn.* **2020**, *39*, 4547–4554. [CrossRef] [PubMed]
28. Karami, Z.; Knoop, B.T.; Dofferhoff, A.S.M.; Blaauw, M.J.T.; Janssen, N.A.; van Apeldoorn, M.; Kerckhoffs, A.P.M.; van de Maat, J.S.; Hoogerwerf, J.J.; Oever, J.T. Few bacterial co-infections but frequent empiric antibiotic use in the early phase of hospitalized patients with COVID-19: Results from a multicentre retrospective cohort study in The Netherlands. *Infect. Dis.* **2020**, *53*, 102–110. [CrossRef]

29. Kearns, F.L.; Sandoval, D.R.; Casalino, L.; Clausen, T.M.; Rosenfeld, M.A.; Spliid, C.B.; Amaro, R.E.; Esko, J.D. Spike-heparan sulfate interactions in SARS-CoV-2 infection. *Curr. Opin. Struct. Biol.* **2022**, *76*, 102439. [CrossRef] [PubMed]
30. Rawson, T.M.; Ming, D.; Ahmad, R.; Moore, L.S.P.; Holmes, A.H. Antimicrobial use, drug-resistant infections and COVID-19. *Nat. Rev. Microbiol.* **2020**, *18*, 409–410. [CrossRef]
31. Hsu, J. How COVID-19 is accelerating the threat of antimicrobial resistance. *BMJ* **2020**, *369*, m1983. [CrossRef]
32. Makowska, M.; Boguszewski, R.; Nowakowski, M.; Podkowińska, M. Self-Medication-Related Behaviors and Poland's COVID-19 Lockdown. *Int. J. Environ. Res. Public Health* **2020**, *17*, 8344. [CrossRef]
33. Tekeba, A.; Ayele, Y.; Negash, B.; Gashaw, T. Extent of and Factors Associated with Self-Medication among Clients Visiting Community Pharmacies in the Era of COVID-19: Does It Relieve the Possible Impact of the Pandemic on the Health-Care System? *Risk Manag. Health Policy* **2021**, *14*, 4939–4951. [CrossRef]
34. Zhang, A.; Hobman, E.; De Barro, P.; Young, A.; Carter, D.; Byrne, M. Self-Medication with Antibiotics for Protection against COVID-19: The Role of Psychological Distress, Knowledge of, and Experiences with Antibiotics. *Antibiotics* **2021**, *10*, 232. [CrossRef]
35. Usman, M.; Farooq, M.; Hanna, K. Environmental side effects of the injudicious use of antimicrobials in the era of COVID-19. *Sci. Total. Environ.* **2020**, *745*, 141053. [CrossRef]
36. Heydargoy, M.H. The Effect of the Prevalence of COVID-19 on Arbitrary Use of Antibiotics. *Iran. J. Med. Microbiol.* **2020**, *14*, 374–378. [CrossRef]
37. Garcia-Vidal, C.; Sanjuan, G.; Moreno-García, E.; Puerta-Alcalde, P.; Garcia-Pouton, N.; Chumbita, M.; Fernandez-Pittol, M.; Pitart, C.; Inciarte, A.; Bodro, M.; et al. Incidence of co-infections and superinfections in hospitalized patients with COVID-19: A retrospective cohort study. *Clin. Microbiol. Infect.* **2020**, *27*, 83–88. [CrossRef]
38. Hughes, S.; Troise, O.; Donaldson, H.; Mughal, N.; Moore, L.S.P. Bacterial and fungal coinfection among hospitalized patients with COVID-19: A retrospective cohort study in a UK secondary-care setting. *Clin. Microbiol. Infect.* **2020**, *26*, 1395–1399. [CrossRef]
39. Iacobucci, G. COVID-19: Risk of death more than doubled in people who also had flu, English data show. *BMJ* **2020**, *370*. [CrossRef]
40. Lv, Z.; Cheng, S.; Le, J.; Huang, J.; Feng, L.; Zhang, B.; Li, Y. Clinical characteristics and co-infections of 354 hospitalized patients with COVID-19 in Wuhan, China: A retrospective cohort study. *Microbes Infect.* **2020**, *22*, 195–199. [CrossRef]
41. Sharifipour, E.; Shams, S.; Esmkhani, M.; Khodadadi, J.; Fotouhi-Ardakani, R.; Koohpaei, A.; Doosti, Z.; Golzari, S.E. Evaluation of bacterial co-infections of the respiratory tract in COVID-19 patients admitted to ICU. *BMC Infect. Dis.* **2020**, *20*, 1–7. [CrossRef]
42. Contou, D.; Claudinon, A.; Pajot, O.; Micaëlo, M.; Flandre, P.L.; Dubert, M.; Cally, R.; Logre, E.; Fraissé, M.; Mentec, H.; et al. Bacterial and viral co-infections in patients with severe SARS-CoV-2 pneumonia admitted to a French ICU. *Ann. Intensiv. Care* **2020**, *10*, 1–9. [CrossRef]
43. Abdoli, A. Helminths and COVID-19 Co-Infections: A Neglected Critical Challenge. *ACS Pharmacol. Transl. Sci.* **2020**, *3*, 1039–1041. [CrossRef]
44. Chen, X.; Liao, B.; Cheng, L.; Peng, X.; Xu, X.; Li, Y.; Hu, T.; Li, J.; Zhou, X.; Ren, B. The microbial coinfection in COVID-19. *Appl. Microbiol. Biotechnol.* **2020**, *104*, 1–9. [CrossRef] [PubMed]
45. Langford, B.J.; So, M.; Raybardhan, S.; Leung, V.; Westwood, D.; MacFadden, D.R.; Soucy, J.-P.R.; Daneman, N. Bacterial co-infection and secondary infection in patients with COVID-19: A living rapid review and meta-analysis. *Clin. Microbiol. Infect.* **2020**, *26*, 1622–1629. [CrossRef] [PubMed]
46. Iskandar, K.; Molinier, L.; Hallit, S.; Sartelli, M.; Hardcastle, T.C.; Haque, M.; Lugova, H.; Dhingra, S.; Sharma, P.; Islam, S.; et al. Surveillance of antimicrobial resistance in low- and middle-income countries: A scattered picture. *Antimicrob. Resist. Infect. Control.* **2021**, *10*, 1–19. [CrossRef] [PubMed]
47. Huttner, B.D.; Catho, G.; Pano-Pardo, J.R.; Pulcini, C.; Schouten, J. COVID-19: Don't neglect antimicrobial stewardship principles. *Clin. Microbiol. Infect.* **2020**, *26*, 808–810. [CrossRef]
48. Cohen, J.F.; Deeks, J.J.; Hooft, L.; Salameh, J.-P.; A Korevaar, D.; Gatsonis, C.; Hopewell, S.; A Hunt, H.; Hyde, C.J.; Leeflang, M.M.; et al. Preferred reporting items for journal and conference abstracts of systematic reviews and meta-analyses of diagnostic test accuracy studies (PRISMA-DTA for Abstracts): Checklist, explanation, and elaboration. *BMJ* **2021**, *372*, n265. [CrossRef]
49. Tricco, A.C.; Lillie, E.; Zarin, W.; O'Brien, K.K.; Colquhoun, H.; Levac, D.; Moher, D.; Peters, M.D.J.; Horsley, T.; Weeks, L.; et al. PRISMA Extension for Scoping Reviews (PRISMA-ScR): Checklist and Explanation. *Ann. Intern. Med.* **2018**, *169*, 467–473. [CrossRef]
50. Ansari, S.; Hays, J.P.; Kemp, A.; Okechukwu, R.; Murugaiyan, J.; Ekwanzala, M.D.; Alvarez, M.J.R.; Paul-Satyaseela, M.; Iwu, C.D.; Balleste-Delpierre, C.; et al. The potential impact of the COVID-19 pandemic on global antimicrobial and biocide resistance: An AMR Insights global perspective. *JAC-Antimicrobial Resist.* **2021**, *3*, dlab038. [CrossRef]

antibiotics

MDPI

Perspective

Beyond Guidelines and Reports on Bacterial Co-/Superinfections in the Context of COVID-19: Why Uniformity Matters

Johan Van Laethem [1,*], Denis Piérard [2] and Sabine D. Allard [1]

1 Department of Internal Medicine, Universitair Ziekenhuis Brussel (UZ Brussel), Vrije Universiteit Brussel (VUB), 1090 Brussels, Belgium
2 Microbiology Department, Universitair Ziekenhuis Brussel (UZ Brussel), 1090 Brussels, Belgium
* Correspondence: johan.vanlaethem@uzbrussel.be

Abstract: Background: In the period following the declaration of the COVID-19 pandemic, more evidence became available on the epidemiology of bacterial co-/superinfections (bCSs) in hospitalized COVID-19 patients. Various European therapeutic guidelines were published, including guidance on rational antibiotic use. Methods: In this letter to the editor, we provide an overview of the largest meta-analyses or prospective studies reporting on bCS rates in COVID-19 patients and discuss why the reader should interpret the results of those reports with care. Moreover, we compare different national and international COVID-19 therapeutic guidelines from countries of the European Union. Specific attention is paid to guidance dedicated to rational antibiotic use. Results: We found a significant heterogeneity in studies reporting on the epidemiology of bCSs in COVID-19 patients. Moreover, European national and international guidelines differ strongly from each other, especially with regard to the content and extent of antibiotic guidance in hospitalized COVID-19 patients. Conclusion: A standardized way of reporting on bCSs and uniform European guidelines on rational antibiotic use in COVID-19 patients are crucial for antimicrobial stewardship teams to halt unnecessary antibiotic use in the COVID-19 setting.

Keywords: bacterial co-infection; bacterial superinfection; antibiotic stewardship; COVID-19

Citation: Van Laethem, J.; Piérard, D.; Allard, S.D. Beyond Guidelines and Reports on Bacterial Co-/Superinfections in the Context of COVID-19: Why Uniformity Matters. *Antibiotics* **2022**, *11*, 1446. https://doi.org/10.3390/antibiotics11101446

Academic Editor: Marcello Covino

Received: 12 September 2022
Accepted: 18 October 2022
Published: 20 October 2022

1. Introduction

The emergence of multidrug-resistant (MDR) bacterial infections has resulted in scarifying future projections. A report from the World Health Organization (WHO) labeled the problem as "so serious, that it threatens the achievements of modern medicine" [1]. For the last few decades, local, national and international efforts have been endorsed by scientific and public health organizations, governments and caregivers to halt the emergence of antimicrobial resistance (AMR). Amongst other interventions, antimicrobial stewardship (AST) teams became the standard of care, guidelines and scientific publications encouraging rational antibiotic use were published, national antibiotic action plans were launched and awareness was cultivated in the general population.

In this perspective article, we first provide a summary of the impact of the COVID-19 pandemic on antimicrobial stewardship efforts, AMR and the growing knowledge on bacterial co-/superinfection epidemiology and antibiotic (over)use in this context. Second, we offer a critical analysis of the major papers reporting on bacterial co-/superinfection (bCS) rates in COVID-19 patients. Last, we discuss the variation in European guidelines for the diagnosis/treatment of these bCSs.

2. Antimicrobial Stewardship Applied to COVID-19 Patients: The Pursuit of Knowledge

At the time the COVID-19 pandemic emerged, AST teams and other actors within the healthcare system constrainedly invested great amounts of time and resources in the contention of the pandemic, the procuration of protective equipment and the reorganization of the healthcare system. The latter inevitably resulted in less stringent antimicrobial stewardship, leaving the battle against AMR in the background. Currently, there is insufficient evidence that the COVID-19 pandemic fueled the AMR threat, as present reports are context-specific and differ geographically. However, there are some ominous signs of increased AMR since the emergence of the COVID-19 pandemic. For the European Union the European Antimicrobial Resistance Surveillance (EARS-Net) network reported a significant rise in carbapenem-resistant *Enterobacterales*, *Pseudomonas aeruginosa* and *Acinetobacter* species as well as vancomycin-resistant enterococci for the year 2020 [2]. Moreover, there was a significant rise in carbapenem use in that same year. The rise in MDR pathogens was most marked in the intensive care setting [3].

In the period following the declaration of the COVID-19 pandemic by the WHO on 11 March 2020, many admitted COVID-19 patients empirically received antibiotics [3,4]. This period of "antibiotic anarchy" can (partly) be explained by the lack of knowledge and lack of guidelines concerning the epidemiology and treatment of presumed bCSs in the context of COVID-19. From May 2020 onwards, the first reports and meta-analyses regarding the incidence and prevalence of bCSs showed very low rates of bacterial co-infections (2.2–8%) and low rates of bacterial superinfections (2.2–20%) in admitted COVID-19 patients. This was in contrast to disproportionally high antibiotic prescribing rates (up to 85%) [4–8]. Soon after the first published reports on bCS incidence, guidelines on the management of presumed bCSs in COVID-19 patients were published by the European Society of Clinical Microbiology and Infectious Disease (ESCMID) (in April 2020) and the WHO (May 2020) [9,10]. Later on, evidence on the good negative predictive value of procalcitonin in excluding bCSs in the context of COVID-19 became available [11,12]. Consequently, over time, a learning effect and a decrease in antibiotic use were noted in some studies [13]. However, antibiotic overprescribing and low quality of antibiotic prescriptions are still prevalent in admitted COVID-19 patients [14].

3. Evidence on bCS Rates in COVID-19 Patients: A Critical Point of View

Robust scientific evidence on COVID-19-related bCS epidemiology and clear therapeutic guidance regarding antibiotic use in COVID-19 patients are crucial for antimicrobial stewardship (AST) teams to prevent antibiotic overprescribing. Despite all progress made to gain expertise on bCS prevalence and antibiotic prescribing patterns in COVID-19 patients, significant knowledge gaps and flaws prevail. Moreover, practical guidance on judicious antibiotic use in COVID-19 patients should be improved, as these guidelines are very heterogeneous and lack specificity. Finally yet importantly, meta-analyses and prospective studies reporting on bCS rates in COVID-19 patients (Table 1) should be interpreted with care for several reasons.

Table 1. Overview of largest (n > 3000) meta-analyses or prospective studies reporting on bacterial co-/superinfection rates in COVID-19 patients (see Supplementary Material S1 for the search strategy).

Reference and Type of Study	Co-/Superinfection * Definitions	Used Diagnostic Criteria	Reported Pathogens	Reported Infections	Setting (Ward/ICU)	Age Group	Co-/Superinfection Rate	Antibiotic Prescription Rate
Langford et al. (2020) [4] Systematic meta-analysis	co-infection: "on presentation" Superinfection: "emerging during the course of illness or during hospitalization"	Not mentioned if clinical and/or microbiological diagnosis	Bacterial	Respiratory tract infections and bloodstream infections	Ward and ICU	Pediatric and adult patients (25%/75%)	co-infection 3.5% superinfection 14.3%	72%

Table 1. *Cont.*

Reference and Type of Study	Co-/Superinfection * Definitions	Used Diagnostic Criteria	Reported Pathogens	Reported Infections	Setting (Ward/ICU)	Age Group	Co-/Superinfection Rate	Antibiotic Prescription Rate
Langford et al. (2022) [5] Systematic meta-analysis	co-infection: not defined	Microbiological diagnosis Exclusion of "presumed" or "suspected" bacterial infection	Bacterial	Respiratory tract infections and bloodstream infections	Ward and ICU	Pediatric and adult patients	co-infection 5.1% secondary infection 13.1%	75%
Lansburry et al. [6] Systematic meta-analysis	co-infection: not defined. Unclear if this term was used to group "co-infections" and "superinfections"	Microbiological diagnosis (culture and PCR)	Bacterial, viral, fungal	Respiratory tract infections and bloodstream infections	Ward and ICU	Pediatric and adult patients	co-infection 7% (bacterial)	NR
Musuuza et al. [7] Systematic meta-analysis	co-infection: "at the time of a SARS-CoV-2 infection" superinfection: "during care for SARS-CoV-2 infection"	Microbiological diagnosis	Bacterial, viral, fungal	Respiratory tract infections	Ward and ICU	Pediatric and adult patients	co-infection 8% superinfection 20% (bacterial)	NR
Russell et al. [8] Original paper	Co-infection: clinically significant positive results from samples collected within 2 days of admission Superinfection: infection occurring > 2 days after hospital admission	Microbiological diagnosis	Bacterial, fungal	Respiratory tract infections and bloodstream infections	Ward and ICU	Not reported	co-infection 0.7% "secondary" infection 1.5%	85%

* Superinfection and secondary infection are used as synonyms. NR: not reported.

First, there is a significant heterogeneity in the used definitions of co-infection and superinfection (also referred to as "secondary infection"). Certain reports use the term co-infection as "every infection contracted before or during the first 48 h of admission", while others use 24 h of admission as a time limit to differentiate co-infection from superinfection. Some even refer to "every infection diagnosed on presentation", while in the meta-analysis of Langford et al. (2022) [5], no definition of co-infection is mentioned. Depending on the used definition, bacterial co-infection and superinfection rates can thus be under- or overestimated. Moreover, international COVID-19 therapeutic guidelines such as the World Health Organization and the European Centre for Disease Control and Prevention guidelines do not define bacterial co-infection and superinfection [10]. Second, although most studies exclusively included microbiological diagnoses, it is not always clear if clinical diagnoses, based on other criteria than microbiological documentation, were included. While the first meta-analysis of Langford et al. (2020) [4] did not mention if included diagnoses were exclusively based on microbiological criteria, their second meta-analysis stated that "presumed" or "suspected" diagnoses of infection were excluded. Depending on the used definition of "infection", the final bCS rate will be different. Third, there is also an important heterogeneity regarding the included microbiological diagnoses. For example, the meta-analysis of Lansburry et al. reports high rates of *Mycoplasma pneumoniae* infections (representing 42% of all reported bacterial infections) [6]. Although no information is provided on the used diagnostic methods, this could be an overestimation due to the inclusion of patients with aspecific serological results. The ISARIC study did not include any *Mycoplasma pneumoniae* infection, as routine testing for atypical pathogens was discontinued in most United Kingdom laboratories during the study period. Fourth, the proportion of included patients depending on the setting (ward versus ICU) and age (adult versus pediatric patients) is not always clearly mentioned. Fifth, some reports include viral and fungal co-/superinfections together with bCSs. Last, the reported infection sites can differ from one study to another. Although most studies focused on both respiratory and bloodstream infections, some also included urinary tract infections whereas others exclusively included respiratory tract co-/superinfections. One should thus pay attention to the reported endpoint when comparing different studies.

4. European Therapeutic COVID-19 Guidelines: An Emphasis on Antibiotic Guidance

As the COVID-19 pandemic progressed, knowledge about bCS epidemiology and antibiotic prescribing in COVID-19 patients increased rapidly. This led to the publication of various antibiotic guidance guidelines. However, studies reporting on bCS rates and guidelines regarding judicious antibiotic use show a significant heterogeneity.

Therefore, we analyzed and compared all published national and international guidelines on COVID-19 therapeutic guidance in the European Union (EU) (see Supplementary Materials for methods, complete data and references). Most countries have published their own national therapeutic guideline (see Figure 1), while others refer to international scientific guidelines, such as those of the World Health Organization [10]. Certain countries, such as Austria, refer to the guidelines of neighboring countries. The majority of the EU countries have also included specific guidance on rational antibiotic use in COVID-19 patients. However, there is a large variability in the extent and content of provided guidance regarding antibiotic use in this setting (Supplementary Material S3, Figure S1). For example, the Dutch Working Party on Antibiotic Policy, as well as the health authorities of Bulgaria, dedicated specific attention to rational antibiotic use in the COVID-19 setting, while in the national guidelines from other countries, such as Belgium, France, Italy, Poland and Spain, only a few sentences on rational AB are found [15–21]. While bacterial co-infections are rare in admitted COVID-19 patients, secondary infections are more prevalent in patients with severe COVID-19. This is probably why most guidelines recommend initiating empiric antibiotics exclusively in patients with severe infection, provided that the need for antibiotics would be regularly evaluated. This is in contrast with the Polish national guidelines, which strongly advise against antibiotics in cases of acute respiratory distress syndrome (ARDS) unless there are evident signs of secondary bacterial infection [20]. Some guidelines limit themselves to recommending antibiotics in cases of suspected bCSs, without elaborating on how to diagnose bCSs [19,20]. The ESCMID guidelines state that only patients with clinical or radiological suspicion of bacterial co-/superinfection should receive empirical antibiotics [9]. However, this is quite vague, as radiological consolidations and clinical signs, such as fever and elevated inflammatory markers, are often present in the context of COVID-19. Therefore, the ECDC guidelines advocate for more clarity in defining secondary bacterial infections in COVID-19 patients [21]. Although the Croatian guidelines advocate that bacterial infection is likely in case of leukocytosis and/or a neutrophil left shift with increased procalcitonin concentration and very high CRP and IL-6 levels, elevated procalcitonin and IL-6 levels have low positive predictive value for bacterial infection in the COVID-19 setting and are also observed in the context of COVID-19 sepsis. Those same guidelines, together with the Danish guidelines, suggest following the "sepsis campaign" guidelines in cases of COVID-19 sepsis [22,23]. However, as COVID-19 sepsis is due to a hyperinflammatory state with a potential cytokine storm, this does not necessarily reflect bacterial sepsis. Yet most sepsis campaign guidelines focus on bacterial sepsis, and this includes the empiric use of antibiotics. Nevertheless, one could agree to empirically start antibiotics in severe and degrading presentations of COVID-19. While the German COVID-19 guidelines recommend antibiotic prescribing at admission in the intensive care unit, those same guidelines paradoxically state that there is no place for prophylactic antibiotics [24,25]. Despite the good negative predictive value of low procalcitonin levels for bCSs [11,12], the use of this predictor is only incorporated in the Latvian guidelines [26].

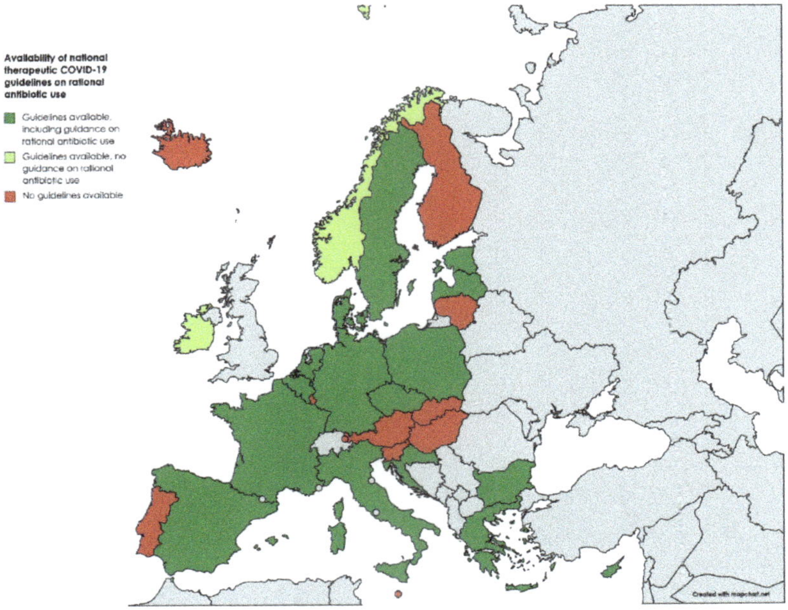

Figure 1. Availability of national therapeutic COVID-19 guidelines on rational antibiotic use. Created with Mapchart: https://www.mapchart.net/europe.html; accessed on 10 October 2022.

5. Conclusions

In conclusion, studies reporting on bCS rates and guidelines regarding judicious antibiotic use show significant heterogeneity. Antibiotic prescribing guidelines depend too much on clinical judgment and should instead take variables into account that have proven to be good predictors or excluders of bCSs, such as procalcitonin. The roles of other potential markers and predictors of bCSs, such as certain comorbidities, the presence of immune suppression and the presence of dense radiological consolidations, are still unclear and should be further investigated.

A standardized way of reporting on bCSs in the context of COVID-19 is the only way to obtain more robust and precise evidence on their incidence and associated risk factors. We therefore strongly advocate for the implementation of international diagnostic guidelines, using predictors and excluders of bCSs and standardized definitions of bCSs. These definitions and guidelines should be dynamic and more detailed. Guidelines should be based on different clinical situations and could indicate the level of diagnostic certainty, and they should evolve according to the best available evidence.

Supplementary Materials: The following supporting information can be downloaded at: https://www.mdpi.com/article/10.3390/antibiotics11101446/s1, Supplementary Material S1: Methods. Supplementary Material S2, Table S1: National and international guidelines on COVID-19 therapeutic guidance in the European Union (EU). Supplementary Material S3, Figure S1: National therapeutic COVID-19 guidelines: number of words dedicated to rational antibiotic use per country. Refs. [9,14–34] cited in Supplementary Materials.

Author Contributions: Conceptualization, J.V.L.; methodology, J.V.L.; investigation, J.V.L.; writing—original draft preparation, J.V.L., S.D.A. and D.P.; writing—review and editing, J.V.L., S.D.A. and D.P.; supervision, D.P. and S.D.A. All authors have read and agreed to the published version of the manuscript.

Funding: This research received no external funding.

Institutional Review Board Statement: Not applicable.

Informed Consent Statement: Not applicable.

Data Availability Statement: See Supplementary Material for additional data.

Acknowledgments: Special thanks to following persons for having provided their guidelines or having answered our requests to collect different guidelines: Mary Vaarpu (Finland), Hanna Rätsep (Estonia), Jorge Ruivo and Nuno Cerca (Portugal), Vladimir Krajinoviç and Neven Papic (Croatia), Ivan Ivanov and Ivailo Alexiev (Bulgaria), L. Viksna (Latvia), Oana Sandulescu and Hristea Adriana (Romania), Sotirios Tsiodras and Spyridon Pournaras (Greece), Guenter Weiss (Austria), Jakub Hurych (Czech Republic), George Petrikkos (Cyprus), Tadeja Matos (Slovenia), Stefan Tyski (Poland), Theodoros Kalpakos (for the translation of the Greek guidelines).

Conflicts of Interest: The authors declare no conflict of interest.

References

1. World Health Organization. AAntimicrobial Resistance. Global Report on Surveillance. 2014. Available online: https://www.who.int/publications/i/item/9789241564748 (accessed on 27 August 2022).
2. European Centre for Disease Prevention and Control. *Antimicrobial Consumption in the EU/EEA (ESAC-Net)—Annual Epidemiological Report 2020*; ECDC: Stockholm, Sweden, 2021.
3. Rawson, T.M.; Moore, L.S.P.; Zhu, N.; Ranganathan, N.; Skolimowska, K.; Gilchrist, M.; Satta, G.; Cooke, G.; Holmes, A.H. Bacterial and Fungal Coinfection in Individuals with Coronavirus: A Rapid Review to Support COVID-19 Antimicrobial Prescribing. *Clin. Infect. Dis.* **2020**, *71*, 2459–2468. [CrossRef] [PubMed]
4. Langford, B.J.; So, M.; Raybardhan, S.; Leung, V.; Westwood, D.; MacFadden, D.R.; Soucy, J.-P.R.; Daneman, N. Bacterial co-infection and secondary infection in patients with COVID-19: A living rapid review and meta-analysis. *Clin. Microbiol. Infect.* **2020**, *26*, 1622–1629. [CrossRef]
5. Langford, B.J.; So, M.; Leung, V.; Raybardhan, S.; Lo, J.; Kan, T.; Leung, F.; Westwood, D.; Daneman, N.; MacFadden, D.R.; et al. Predictors and microbiology of respiratory and bloodstream bacterial infection in patients with COVID-19: Living rapid review update and meta-regression. *Clin. Microbiol. Infect.* **2022**, *28*, 491–501. [CrossRef] [PubMed]
6. Lansbury, L.; Lim, B.; Baskaran, V.; Lim, W.S. Co-infections in people with COVID-19: A systematic review and meta-analysis. *J. Infect.* **2020**, *81*, 266–275. [CrossRef]
7. Musuuza, J.S.; Watson, L.; Parmasad, V.; Putman-Buehler, N.; Christensen, L.; Safdar, N. Prevalence and outcomes of co-infection and superinfection with SARS-CoV-2 and other pathogens: A systematic review and meta-analysis. *PLoS ONE* **2021**, *16*, e0251170. [CrossRef]
8. Russell, C.D.; Fairfield, C.J.; Drake, T.M.; Turtle, L.; Seaton, R.A.; Wootton, D.G.; Sigfrid, L.; Harrison, E.M.; Docherty, A.B.; I de Silva, T.; et al. ISARIC4C investigators. Co-infections, secondary infections, and antimicrobial use in patients hospitalised with COVID-19 during the first pandemic wave from the ISARIC WHO CCP-UK study: A multicentre, prospective cohort study. *Lancet Microbe* **2021**, *2*, e354–e365. [CrossRef]
9. Bartoletti, M.; Azap, O.; Barac, A.; Bussini, L.; Ergonul, O.; Krause, R.; Paño-Pardo, J.R.; Power, N.R.; Sibani, M.; Szabo, B.G.; et al. ESCMID COVID-19 living guidelines: Drug treatment and clinical management. *Clin. Microbiol. Infect.* **2022**, *28*, 222–238. [CrossRef] [PubMed]
10. World Health Organization. *COVID-19 Clinical Management: Living Guidance*; 25 January 2021; Report No.: WHO/2019-nCoV/clinical/2021.1; WHO: Geneva, Switzerland, 2021.
11. Dolci, A.; Robbiano, C.; Aloisio, E.; Chibireva, M.; Serafini, L.; Falvella, F.S.; Pasqualetti, S.; Panteghini, M. Searching for a role of procalcitonin determination in COVID-19: A study on a selected cohort of hospitalized patients. *Clin. Chem. Lab. Med.* **2020**, *59*, 433–440. [CrossRef]
12. Vaughn, V.M.; Gandhi, T.N.; Petty, L.A.; Patel, P.K.; Prescott, H.C.; Malani, A.N.; Ratz, D.; McLaughlin, E.; Chopra, V.; Flanders, S.A. Empiric Antibacterial Therapy and Community-onset Bacterial Coinfection in Patients Hospitalized with Coronavirus Disease 2019 (COVID-19): A Multi-hospital Cohort Study. *Clin. Infect. Dis.* **2021**, *72*, e533–e541. [CrossRef] [PubMed]
13. Van Laethem, J.; Wuyts, S.C.M.; Pierreux, J.; Seyler, L.; Verschelden, G.; Depondt, T.; Meuwissen, A.; Lacor, P.; Piérard, D.; Allard, S.D. Presumed Urinary Tract Infection in Patients Admitted with COVID-19: Are We Treating Too Much? *Antibiotics* **2021**, *10*, 1493. [CrossRef]
14. Sieswerda, E.; de Boer, M.G.; Bonten, M.M.; Boersma, W.G.; Jonkers, R.E.; Aleva, R.M.; Kullberg, B.-J.; Schouten, J.A.; van de Garde, E.M.; Verheij, T.J.; et al. Recommendations for antibacterial therapy in adults with COVID-19—An evidence based guideline. *Clin. Microbiol. Infect.* **2021**, *27*, 61–66. [CrossRef]
15. Recommendations for Antibiotic Treatment of Patients with COVID-19 Infection. Available online: https://www.mh.government.bg/media/filer_public/2021/08/09/preporki_-_mikrobiologiia.pdf (accessed on 25 August 2022).
16. Interim Clinical Guidance for Adults with Confirmed COVID-19 in Belgium, July 2022; Version 29. Available online: https://covid-19.sciensano.be/sites/default/files/Covid19/COVID-19_InterimGuidelines_Treatment_ENG.pdf (accessed on 25 August 2022).
17. Haut Conseil de la Santé Publique. Coronavirus SARS-CoV-2: Recommandations sur L'usage des Anti Infectieux. Available online: https://www.hcsp.fr/Explore.cgi/AvisRapportsDomaine?clefr=849 (accessed on 25 August 2022).

18. Italian Society of Infectious and Tropical Diseases. Vademecum for the Care of People with COVID-19 Disease. Edition 2.0, 13 March 2020. Available online: https://www.eahp.eu/sites/default/files/covid19_vademecum_2.0_13_marzo_2020.03_11.pdf (accessed on 25 August 2022).
19. Flisiak, R.; Horban, A.; Jaroszewicz, J. Diagnosis and therapy of SARS-CoV-2 infection: Recommendations of the Polish Association of Epidemiologists and Infectiologists as of 12 November 2021. Annex no. 1 to the Recommendations of 26 April 2021. *Pol. Arch. Intern. Med.* **2021**, *131*, 16140. [CrossRef]
20. Recomendaciones SEIMC Para el Manejo Clínico de Pacientes Con COVID-19. Available online: https://covid19.seimc.org/wp-content/uploads/2022/05/SEIMC-Recomendaciones-COVID_24-mayo-2022.pdf (accessed on 25 August 2022).
21. European Centre for Disease Prevention and Control. Treatment and Pharmaceutical Prophylaxis of COVID-19. Available online: https://www.ecdc.europa.eu/en/covid-19/latest-evidence/treatment (accessed on 25 August 2022).
22. Ministry of Health of the Republic of Croatia. Guidelines for the Treatment of Patients with Coronavirus Disease 2019 (COVID-19). Version 5 of 8 February 2022. Available online: https://www.koronavirus.hr/uploads/Smjernice_za_lijecenje_oboljelih_od_koronavirusne_bolesti_2019_COVID_19_verzija_5_od_08_veljace_2022_edb7d62da1.pdf (accessed on 25 August 2022).
23. Jeschke, K.N.; Bonnesen, B.; Hansen, E.F.; Jensen, J.-U.S.; Lapperre, T.S.; Weinreich, U.M.; Hilberg, O. Guideline for the management of COVID-19 patients during hospital admission in a non-intensive care setting. *Eur. Clin. Respir. J.* **2020**, *7*, 1761677. [CrossRef] [PubMed]
24. Malin, J.J.; Spinner, C.D.; Janssens, U.; Welte, T.; Weber-Carstens, S.; Schälte, G.; Gastmeier, P.; Langer, F.; Wepler, M.; Westhoff, M.; et al. Key pharmacologic recommendations from a national German living guideline using an Evidence to Decision Framework (last updated 17 May 2021). *Infection* **2022**, *50*, 93–106. [CrossRef]
25. Kluge, S.; Janssens, U.; Welte, T.; Weber-Carstens, S.; Marx, G.; Karagiannidis, C. Empfehlungen zur intensivmedizinischen Therapie von Patienten mit COVID-19. *Med. Klin. Intensivmed. Notfmed.* **2020**, *115*, 175–177. [CrossRef] [PubMed]
26. Disease Prevention and Control Centre of Latvia (SPKC). RECOMMENDATIONS. SARS-CoV-2 Infection and Epidemiology of COVID-19, Diagnostics, Clinical Developments and Problems. March 2021. Available online: https://www.spkc.gov.lv/sites/spkc/files/media_file/covid_19_rekomendacijas_marts-gatavs-1.pdf (accessed on 25 August 2022).
27. Štefan, M.; Chrdle, A.; Husa, P.; Beneš, J.; Dlouhý, P. COVID-19: Diagnosis and treatment. *Klin Mikrobiol Infekc Lek.* **2021**, *27*, 61–87. [PubMed]
28. Hospital manual treatment of infection. COVID-19. November 2020. Not available online. To access the pdf file, please contact the corresponding author of this paper.
29. Kallaste, A.; Härma, E.; Rätsep, H. COVID-19 patsiendi käsitlusjuhend Tartu Ülikooli Kliinikumis ja Põhja-Eesti Regionaalhaiglas. Version 1. Revised version: 11.11.2020. To access the pdf file, please contact the corresponding author of this paper.
30. Kluge, S.; Janssens, U.; Spinner, C.D.; Pfeifer, M.; Marx, G.; Karagiannidis, C.; Guideline, group. Clinical practice guideline: Recommendations on in-hospital treatment of patients with COVID-19. *Dtsch. Arztebl. Int.* **2021**, *118*, 1–7.
31. Greece: Hellenic Society of Infectious Diseases. Therapeutic Algorithm of Adult Hospitalized Patients with COVID-19*. February 2022. Available online: https://eody.gov.gr/wp-content/uploads/2022/02/covid_19_algorithmos-nosileuomenon_20220217.pdf (accessed on 10 September 2022).
32. HSE Interim Guidance for the Pharmacological Management of Patients with COVID-19. Available online: https://hse-ie.libguides.com/ld.php?content_id=33534048 (accessed on 25 August 2022).
33. Norwegian Institute of Public Health. NIPH Systematic and Living Map on COVID-19 Evidence. Available online: https://www.nornesk.no/forskningskart/NIPH_interventionsTreatMap.html (accessed on 25 August 2022).
34. Nationellt Vårdprogram för Misstänkt Och Bekräftad COVID-19. Version 4 maj 2022. Framtaget av Svenska Infektionsläkarföreningen, Svenska Hygienläkarföreningen och Föreningen för Klinisk Mikrobiologi. Available online: https://infektion.net/wp-content/uploads/2022/09/nationellt-vardprogram-covid-version-4-1.pdf (accessed on 30 August 2022).

antibiotics

MDPI

Article

Mid-Regional Pro-Adrenomedullin as a Prognostic Factor for Severe COVID-19 ARDS

Etienne de Montmollin [1,2], Katell Peoc'h [3,4], Mehdi Marzouk [2], Stéphane Ruckly [1], Paul-Henri Wicky [2], Juliette Patrier [2], Pierre Jaquet [2], Romain Sonneville [1,2], Lila Bouadma [1,2] and Jean-François Timsit [1,2,*]

1 Université de Paris Cité, INSERM, UMR 1137, IAME, 75018 Paris, France
2 Medical and Infectious Diseases Intensive Care Unit, APHP, Bichat-Claude Bernard Hospital, 75018 Paris, France
3 Université de Paris Cité, INSERM, UMRs 1149, CRI, 75018 Paris, France
4 Biochemistry Department, APHP, Bichat-Claude Bernard Hospital, 75018 Paris, France
* Correspondence: jean-francois.timsit@aphp.fr

Abstract: Mid-regional proadrenomedullin (MR-proADM) protects against endothelial permeability and has been associated with prognosis in bacterial sepsis. As endothelial dysfunction is central in the pathophysiology of severe SARS-CoV-2 infection, we sought to evaluate MR-proADM both as a prognostic biomarker and as a marker of bacterial superinfection. Consecutive patients admitted to the ICU for severe SARS-CoV-2 pneumonia were prospectively included and serum was bio-banked on days 1, 3, and 7. MR-proADM levels were measured blindly from clinical outcomes in batches at the end of follow-up. Among the 135 patients included between April 2020 and May 2021, 46 (34.1%) had died at day 60. MR-proADM levels on days 1, 3, and 7 were significantly higher in day-60 non-survivors. The area under the curve (AUC) of the receiver operating characteristic (ROC) curve (0.744, $p < 0.001$) of day-1 MR-proADM compared favorably with the AUC ROC curve of day-1 procalcitonin (0.691, $p < 0.001$). Serial MR-proADM measurements on days 3 and 7 may add prognostic information. After adjusting for CRP, LDH, and lymphocyte values, day-1 MR-proADM remained significantly associated with day-60 mortality. MR-proADM concentrations were significantly higher in patients with respiratory superinfections (on days 3 and 7) and bloodstream infections (on days 1, 3, and 7) than in patients without infection. Our results suggest that MR-proADM is a good predictor of outcome in severe SARS-CoV-2 infection and could be a useful tool to assess bacterial superinfection in COVID-19 patients.

Keywords: Mid-regional Pro-adrenomedullin; COVID-19; SARS-CoV-2; ARDS; biomarker; prognosis; superinfection

Citation: de Montmollin, E.; Peoc'h, K.; Marzouk, M.; Ruckly, S.; Wicky, P.-H.; Patrier, J.; Jaquet, P.; Sonneville, R.; Bouadma, L.; Timsit, J.-F. Mid-Regional Pro-Adrenomedullin as a Prognostic Factor for Severe COVID-19 ARDS. *Antibiotics* **2022**, *11*, 1166. https://doi.org/10.3390/antibiotics11091166

Academic Editors: Sabine Danielle Allard and Johan Van Laethem

Received: 31 July 2022
Accepted: 26 August 2022
Published: 29 August 2022

Publisher's Note: MDPI stays neutral with regard to jurisdictional claims in published maps and institutional affiliations.

1. Introduction

Coronavirus Disease (COVID-19) can present with a wide range of clinical severity, from asymptomatic infection to acute respiratory distress syndrome (ARDS) and death. As of August 2022, 591.7 million cumulative cases and 6.5 million deaths due to COVID-19 have been reported to the World Health Organization [1]. Acute hypoxemic respiratory failure is the main reason for intensive care unit (ICU) admission, and critically ill COVID-19 patients show mortality rates up to 40% [2]. The progression from a mild to severe disease is multifactorial but appears mainly driven by significant inflammation and microvascular thrombosis with evidence of endotheliitis [3–5].

Adrenomedullin is an endogenous vasoregulatory peptide that has been shown to play a role in preserving the integrity and stability of the endothelium after severe infection [6]. Mid-regional proadrenomedullin (MR-proADM) is used as a surrogate marker for adrenomedullin, as its levels are directly proportional to adrenomedullin which has a short half-life. Higher levels of MR-proADM in septic critically ill patients have been associated with disease severity [7] and mortality [8,9]. MR-proADM has also shown potential as a risk stratification biomarker for bacterial community-acquired pneumonia [10]. Finally MR-proADM also appears to be a useful diagnostic tool for sepsis [11].

Being a marker of endothelial dysfunction, MR-proADM has been investigated in SARS-CoV-2 infection. Higher levels have been associated with mortality in the general population [12–15] and MR-proADM appeared to be a good risk stratification tool in the specific settings of the Emergency Room [16,17] and the ICU [18–20]. However, studies performed in the ICU setting used small patient samples with high heterogeneity of patient severity [21] and measured outcomes $\leq$ 30 days after ICU admission, which may be insufficient for such a study population.

Severe COVID-19 patients are at increased risk of bacterial superinfections, which contribute to ICU mortality [22,23]. These infections may be difficult to diagnose and lead to antibiotic overuse, which has led to a spread of antimicrobial resistance since the beginning of the pandemic [24]. In this context, MR-proADM could be an interesting biomarker for the diagnosis of bacterial superinfection. Surprisingly, no study evaluated MR-proADM as a diagnostic tool for bacterial sepsis in COVID-19 patients.

In this study, we aimed to evaluate MR-proADM as a prognostic biomarker in critically ill patients with severe SARS-CoV-2 pneumonia, and as a diagnostic tool for bacterial superinfection.

2. Results

2.1. Population Characteristics

Between April 2020 and May 2021, among 1294 admissions to the ICU of Bichat-Claude Bernard university hospital, 358 patients had severe SARS-CoV-2 pneumonia confirmed by polymerase chain reaction (PCR) and 135 patients with systematic bio-banking, and at least one measurement at day 1, 3, or 7 was included in the analysis (Supplementary Figure S1) and followed-up for 39 (13–126) days. Day-1, day-3, and day-7 MR-proADM measurements were available in 120 out of 135 (88.9%), 119 out of 126 (94.4%), and 83 out of 86 (96.5%) patients, respectively. Among the three proADM measurement time points, 69 patients completed three, while 49 patients completed two, and 17 patients completed one.

Included patients were females in 43 (31.9%) cases, with a median age of 62.7 (51.6–71.2) years and an admission Simplified Acute Physiology Score (SAPS) II score of 27 (21–39) (Table 1). ICU admission occurred 9 (7–12) days after symptom onset. On day 1, 11 (8.1%) patients were under veno-venous extracorporeal membrane oxygenation, 19 (14.1%) patients were under invasive mechanical ventilation, 78 (57.8%) had non-invasive oxygenation techniques (non-invasive ventilation, continuous positive airway pressure or high flow nasal oxygen), and 27 (20%) had standard oxygen support. Antiviral treatment consisted of remdesivir for 65 (48.1%) patients. Immunomodulating treatments consisted of steroids in 127 (94.1%) cases, and anti-IL6 (tocilizumab) or anti-IL-1 (anakinra) in three (2.2%) cases. During ICU stay, 59 (44%) patients required invasive mechanical ventilation, 27 (20%) vasopressors, and 30 (22%) renal replacement therapy. ICU, hospital, and day-60 mortality rates were 30.4%, 36.3%, and 34.1%, respectively. ICU and hospital lengths of stay were 10 (6–22) and 16 (10–31) days, respectively.

2.2. Mid-Regional Pro-Adrenomedullin and Day-60 Mortality

Values of MR-proADM and several other biological markers are presented in Table 1. In patients that died before day 60, MR-proADM levels were significantly higher at all time points compared to survivors (Table 1 and Figure 1a). The areas under the receiver operating characteristic curve (AUROC) of MR-proADM for predicting day-60 mortality were 0.74 on day 1, 0.73 on day 3, and 0.74 on day 7 (Figure 1b). When choosing a cut-point of 1 nmol/L for day-1 MR-proADM (median value of the study population), sensitivity and specificity for predicting day-60 mortality were 77.5% (95% confidence interval (CI) 62.5–87.7) and 68.8% (95% CI 57.9–77.9), respectively.

Table 1. Population characteristics at ICU admission.

	All Patients n = 135	Day-60 Survivors n = 89	Day-60 Decedents n = 46	*p*
Demographics and comorbidities				
Age, years	62.7 (51.6–71.2)	58.3 (49.2–66.1)	71.1 (62.3–75.9)	<0.01
Female gender	43 (31.9)	30 (33.7)	13 (28.3)	0.52
Body Mass Index	29 (25.8–34)	29.6 (26.3–34)	28.6 (24.3–31.2)	0.09
Diabetes	40 (29.6)	17 (19.1)	23 (50)	<0.01
Chronic diseases (Knaus $\geq$ 1)	60 (44.4)	36 (40.4)	24 (52.2)	0.19
Immunodepression	16 (11.9)	5 (5.6)	11 (23.9)	<0.01
Characteristics at ICU admission				
SAPS II	27 (21–39)	25 (18–34)	34 (26–48)	<0.01
Time from 1st symptoms to ICU admission, days	9 (7–12)	9 (7–12)	8 (7–15)	0.80
Respiratory SOFA	3 (1–4)	3 (1–4)	3 (2–4)	0.50
Extra-respiratory SOFA	1 (0–4)	1 (0–3)	1 (1–6)	<0.01
Ventilatory status at day 1				0.25
None	27 (20)	19 (21.3)	8 (17.4)	.
NIV/HFNC/CPAP	78 (57.8)	54 (60.7)	24 (52.2)	.
IMV/ECMO	30 (22.2)	16 (18)	14 (30.4)	.
Steroid therapy at day 1	127 (94.1)	85 (95.5)	42 (91.3)	0.44
Laboratory data				
MR-proADM, nmol/L				
Day 1	1 (0.7–1.6)	0.8 (0.7–1.1)	1.4 (1.1–2.7)	<0.01
Day 3	0.9 (0.7–1.7)	0.9 (0.7–1.2)	1.5 (0.9–3.6)	<0.01
Day 7	1 (0.7–1.9)	0.9 (0.6–1.2)	1.4 (1–3.6)	<0.01
IL-6 at day 1, pg/mL	36.3 (9–88)	29.4 (7.6–72)	58.3 (13.2–168)	0.11
CRP at day 1, mg/L	127 (68–177)	129 (57–177)	123.5 (93–172)	0.40
Procalcitonin at day 1, µg/L	0.3 (0.1–1.4)	0.2 (0.1–1)	0.6 (0.3–2)	<0.01
LDH at day 1, UI/L	439 (337–573)	427 (335.5–550)	474 (362–632)	0.16
Lymphocytes at day 1, G/L	0.97 (0.59–1.32)	1.01 (0.63–1.39)	0.85 (0.46–1.12)	0.04
D-dimers at day 1, µg/L	916 (556–1768)	811 (537–1431)	1161 (667–2594)	0.07
Ferritin at day 1, µg/L	851 (397–1668)	804 (344–1448)	1088 (575–2418)	0.03

Abbreviations: ICU, Intensive care Unit; SAPS, Simplified Acute Physiology Score; SOFA, Sequential Organ Failure Assessment; NIV, Non-invasive ventilation; HFNC, High Flow Nasal Canula; CPAP, Continuous Positive Airway Pressure; IMV, Invasive Mechanical Ventilation; ECMO, Extra-Corporeal Membrane Oxygenation, MR-proADM, Mid-regional proadrenomedullin; CRP, C-reactive protein; LDH, Lactate Dehydrogenase.

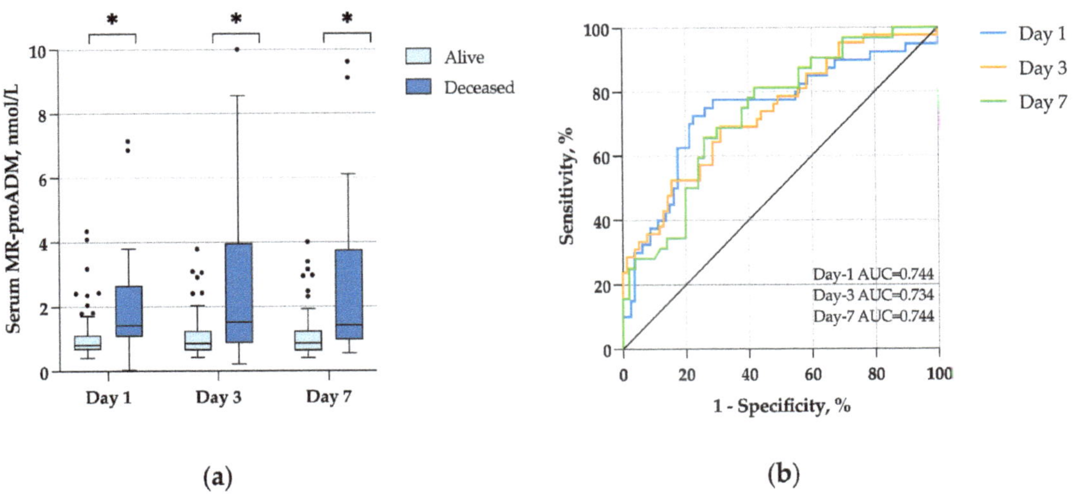

Figure 1. (**a**) MR-proADM concentrations according to day-60 survival; (**b**) ROC curves for day-60 survival of MR-proADM on days 1, 3, and 7. * $p < 0.05$.

Survival curves according to this cut-point are presented in Figure 2. Survival curves according to the same cut-point for day-3 and day-7 MR-proADM are presented in Figure S2 and show a significant difference in day-60 mortality (log-rank test, $p < 0.001$ on day 3 and $p = 0.002$ on day 7). In a landmark analysis on day 3, the delta between day-3 and day-1 MR-proADM was significantly associated with day-60 mortality (HR 1.20, 95% CI 1.01–1.43, $p = 0.04$) (Table S1). On day 7, the delta between day-7 and day-1 MR-proADM concentrations was not associated with day-60 mortality (HR 1.26, 95% CI 0.90–1.76, $p = 0.16$).

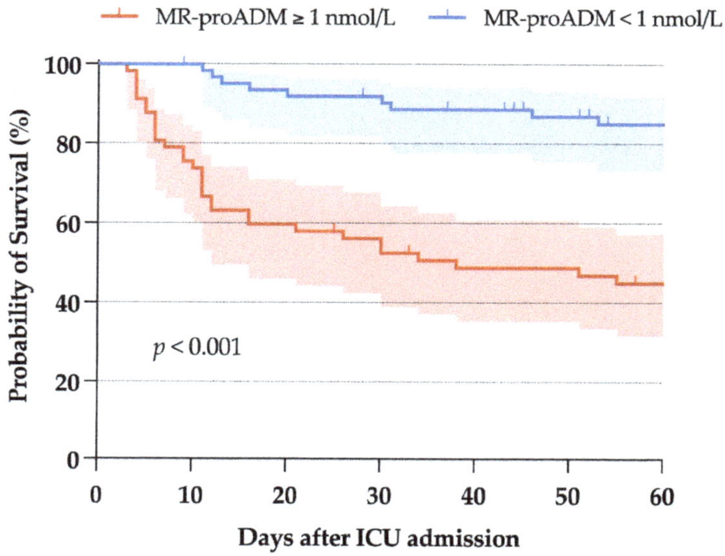

Figure 2. Survival curves according to a cut-point of 1 nmol/L of MR-proADM on day 1.

On day 1, MR-proADM compared favorably to other prognostic biomarkers identified in the literature, AUROC for procalcitonin, ferritin, d-dimers, and C-reactive protein being 0.69, 0.63, 0.60, and 0.55, respectively (Figure 3). It also compared favorably to the Sequential Organ Failure Assessment (SOFA) score on day 1 (AUROC 0.65). The combination of MR-proADM and procalcitonin improved only slightly prognostic accuracy, with an AUROC of 0.76. In multivariate analysis and at all time points, after adjusting for C-reactive protein, lactate dehydrogenase, and lymphocyte count, MR-proADM remained significantly associated with day-60 mortality (Table 2).

2.3. *Mid-Regional Pro-Adrenomedullin and Bacterial Infections*

During ICU stay, 52 (38.5%) patients presented with bacterial nosocomial pneumonia with a delay of 7 (4.5–9) days, of which 36 (26.7%) patients were ventilator-acquired pneumonia. Bacteremia occurred in 34 (25.2%) patients during the same period, with a delay of 9.5 (6–13) days. MR-proADM levels on day 3 (1.2 (0.8–2.1) vs. 0.9 (0.7–1.5), $p < 0.01$) and day 7 (1.2 (0.9–2.5) vs. 0.9 (0.6–1.2), $p < 0.01$) were significantly higher in patients with bacterial pneumonia, but not on day 1 (1.1 (0.7–1.6) vs. 0.8 (0.7–1.5), $p = 0.22$). MR-proADM levels were significantly higher in patients with bacteremia on day 1 (1.3 (1.0–2.4) vs. 0.8 (0.7–1.3), $p < 0.01$), day 3 (1.4 (0.9–2.9) vs. 0.9 (0.7–1.5), $p < 0.01$) and day 7 (1.1 (0.9–2.6) vs. 0.9 (0.7–1.4), $p < 0.01$). MR-proADM levels according to the occurrence of bacteremia or bacterial pneumonia, on days 1, 3, and 7, are presented in Figure 4.

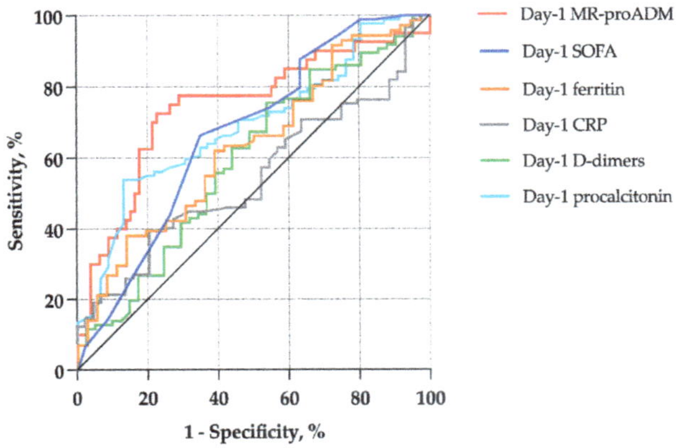

Figure 3. ROC curves of the association of day-60 mortality with various biomarkers on day 1 and the SOFA score. Abbreviations: MR-proADM, Mid-regional proadrenomedullin; SOFA, Sequential Organ Failure Assessment; CRP, C-reactive protein.

Table 2. Adjusted landmark analysis of the association of MR-proADM with day-60 mortality on days 1, 3, and 7.

	Hazard Ratio	95% Confidence Interval	p
Landmark at day 1 (n = 135)			
Day-1 MR-proADM	1.17	(1.06–1.28)	<0.01
Day-1 CRP	1.00	(1–1)	0.50
Day-1 lymphocytes count	1.00	(1–1)	0.28
Day-1 LDH	1.00	(1–1)	<0.01
Landmark at day 3 (n = 135)			
Day-3 MR-proADM	1.17	(1.06–1.28)	<0.01
Day-1 CRP	1.00	(0.99–1)	0.50
Day-1 lymphocytes count	1.00	(1–1)	0.28
Day-1 LDH	1.00	(1–1)	<0.01
Landmark at day 7 (n = 128)			
Day-7 MR-proADM	1.19	(1.1–1.31)	<0.01
Day-1 CRP	1.00	(0.99–1)	0.65
Day-1 lymphocytes count	1.00	(1–1)	0.62
Day-1 LDH	1.00	(1–1)	<0.01

Missing data imputed by multiple imputation. Hazard ratios are computed per one point of each variable. Abbreviations: MR-proADM, Mid-regional proadrenomedullin; CRP, C-reactive protein; LDH, Lactate Dehydrogenase.

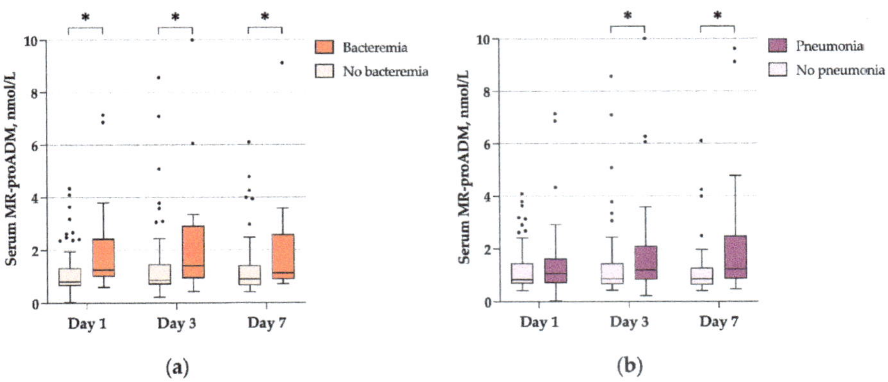

Figure 4. (**a**) MR-proADM concentrations according to the occurrence of pneumonia on days 1, 3, and 7. (**b**) MR-proADM concentrations according to the occurrence of pneumonia on days 1, 3, and 7. Abbreviations: MR-proADM, Mid-regional proadrenomedullin. * $p < 0.05$.

3. Discussion

3.1. Main Findings

Using high-quality prospectively collected data from critically ill COVID-19 patients admitted to a large French COVID-19 reference center, we showed that MR-proADM concentrations were strongly associated with day-60 mortality. The prognostic accuracy of baseline MR-proADM was higher than commonly measured laboratory parameters and the SOFA score. We also showed that the AUROC for the prediction of day-60 mortality remained high on days 3 and 7, but that serial measurements might not be useful at all time points. When choosing a cut-point of 1 nmol/L, sensitivity and specificity for predicting day-60 mortality were 77.5% and 68.8, respectively, with good discrimination of survival curves. When evaluating the predictive accuracy of respiratory bacterial superinfection MR-proADM concentrations were not significantly higher at baseline.

3.2. Interpretation

Regarding mortality risk stratification, our results are in line with previous studies performed in the ICU population showing AUROCs between 0.73 and 0.85 for 28-day mortality, with optimal cut-points between 1 and 1.8 nmol/L [18–20]. We used cut-points according to the MR-proADM distribution and previous studies to avoid overfitting [25]. Thus, the value of 1 nmol/L we chose was close to the median value of our sample and is in the range of published cut-points for mortality. Comparatively, in the general ward [12–15] and the emergency room [16,17] settings, MR-proADM also showed interesting risk stratification capabilities, with risk prediction of ICU admission, need for invasive mechanical ventilation, or death. These results have been confirmed in a pooled analysis of 6 studies and 487 patients, where MR-proADM values were increased by 74% (95% CI 46–103) in COVID-19 patients with critical illness compared to those without [21]. We believe risk stratification to be of paramount importance for COVID-19 patients, as patients with the highest values may benefit most from anti-viral or anti-inflammatory therapies such as steroids, interleukin-6 receptor antagonists, or anti-JAK molecules. In the context of COVID-19 patients, we show that MR-proADM has better prognostic capabilities than procalcitonin. These results are in accordance with published literature regarding the general sepsis population, where MR-proADM appeared to be a prognostic biomarker superior to procalcitonin [26,27].

The analysis of serial measurements of MR-proADM on days 1, 3, and 7 brings valuable information. First, we showed that prognostic accuracy for day-60 mortality is equivalent at each time point, meaning that MR-proADM can be measured at any time during the first few days of ICU admission. Second, the delta between day-3 and day-1 MR-proADM was significantly associated with day-60 mortality, suggesting that MR-proADM could be used to monitor COVID-19 patients. The fact that the delta between day-7 and day-1 MR-proADM was not significantly associated with day-60 mortality may be related to a loss of power due to a smaller patient sample on day 7. In a general population of 89 COVID-19 patients, Gregoriano et al. also found that MR-proADM remained low during the whole follow-up period in survivors, whereas non-survivors had a step-wise increase from baseline [14]. Given the small patient sample, our results warrant further studies to evaluate MR-proADM as a monitoring biomarker of disease progression.

MR-proADM has been proven a performing biomarker for the diagnosis of sepsis, with a calculated AUROC of 0.91 in a recent meta-analysis, and an optimal cut-point value of 1–1.5 nmol/L [11]. It has also shown a good diagnostic accuracy in specific infections, such as complicated urinary tract infections [28] and spontaneous bacterial peritonitis [29]. Thus, we sought to evaluate the diagnostic performance of MR-proADM for bacterial superinfection. Interestingly, we found that MR-proADM concentrations were significantly higher in patients with bacteremia and bacterial pneumonia. Bacterial pneumonia is frequent in severe COVID-19 ARDS, with up to 44% of mechanically ventilated patients developing ventilator-acquired pneumonia [30]. These infections can be difficult to diagnose, as COVID-19 patients may exert persistent systemic inflammation, and chest x-rays may be difficult to analyze due to the underlying viral pneumonia. While at the time of intubation less than 25% of patients present bacterial superinfection, ICU patients are frequently given systematic empiric antibiotic therapy. This strategy has led to an increase in antimicrobial resistance [24], and tools to identify patients with a high probability of bacterial superinfection are dearly needed. As such, we show that MR-proADM could be a useful marker to monitor bacterial superinfection in these patients, but our results need validation in larger cohorts.

3.3. Strenghts and Limitations

The strengths of our study are the prospective design and quality of collected data, including a follow up of 60 days, relevant for severe COVID-19 patients with extensive lengths of stay. The limitations of our study are: (1) a monocentric study design, (2) the small patient sample, despite being the largest published cohort in the literature, (3) the lack of external validation, and (4) the span of the study, covering different COVID-19 waves and SARS-CoV2 variants. Indeed, during each wave, the dominant variant presented distinct clinical and biological characteristics, including inflammatory response profile [31,32]. Hence, the prognostic accuracy of MR-proADM for each variant may have differed, but this has not been evaluated due to an insufficient patient sample.

4. Materials and Methods

4.1. Study Population

From April 2020 to May 2021, we included all adult patients that were admitted to the medical ICU of our hospital for severe SARS-CoV2 pneumonia and had had prospective serum bio-banking in the context of the OUTCOMEREA database. The bio-banking was conducted with the understanding and consent of each participant or surrogate. The OUTCOMEREA database has been approved by the French Advisory Committee for Data Processing in Health Research and the French Informatics and Liberty Commission (CNIL, registration no. 8999262). The database protocol was submitted to the Institutional Review Board of the French society of intensive care (CE-SRLF 22-76) on 12 September 2021. There were no exclusion criteria.

4.2. Data Collection and Definitions

The primary endpoint was the survival rate 60 days after ICU admission, and patients were followed-up to this time point or death. Data were prospectively collected at admission (demographics, chronic diseases, admission features, baseline severity indexes) and daily throughout the ICU stay (specific SARS-CoV2 treatments, need for invasive mechanical ventilation, need for vasopressors, need for renal replacement therapy, bacterial pneumonia and bacteremia, length of stay (LOS) and vital status at ICU and hospital discharge), using an anonymized electronic case report form. Severity of illness was graded at ICU admission with the use of the SAPS II [33] and the SOFA scores [34]. Immunodepression was defined as the use of long-term (>3 months) steroids, use of other immunosuppressant drugs, solid organ transplantation, solid tumor requiring chemotherapy in the last 5 years, hematologic malignancy, or HIV infection.

Serums were systematically collected on days 1, 3, and 7 (unless the patient was discharged from the ICU), allowed to clot at room temperature for 45 min, and then aliquoted and stored at −80 °C until assayed. Day-1, day-3, and day-7 MR-proADM concentrations were then measured blindly from clinical outcomes in batches at the end of follow-up, using an immunological assay with the TRACE technology (B·R·A·H·M·S MR-proADM KRYPTOR assay).

4.3. Statistical Analysis

Quantitative variables are presented as median, 1st, and 3rd quartiles, and compared between groups with the Mann–Whitney test or t-test, as appropriate. Qualitative variables are presented as frequency and percentage and compared with the Chi-square test or

Fisher exact test as appropriate. For the analysis of bacterial superinfections, only episodes occurring after each time point analysis were considered. The association of MR-proADM with day-60 mortality was determined using a Cox proportional hazard model, adjusted on biomarkers associated with the outcome in the literature. Landmark analysis was used to evaluate this association at each time point (day 1, day 3, and day 7). Missing data, when at random, were handled by multiple imputations [35]. All statistical analyses were carried out with SAS 9.4 (SAS Institute Inc., Cary, NC, USA). A p-value of 0.05 and lower was considered statistically significant.

5. Conclusions

Our results suggest that MR-proADM is a promising predictor of outcome in critically ill patients with severe SARS-CoV-2 infection, superior to procalcitonin or the SOFA score. Furthermore, serial measurements may help monitor disease progression. In the event of future COVID-19 waves, MR-proADM could be used both for risk stratification and triage of patients presenting with severe SARS-CoV-2 pneumonia and for monitoring of bacterial superinfection in COVID-19 ICU patients.

Supplementary Materials: The following supporting information can be downloaded at: https://www.mdpi.com/article/10.3390/antibiotics11091166/s1, Figure S1: Study flowchart; Figure S2: Survival curves according to a cut-point of 1 nmol/L of MR-proADM, on days 3 and 7; Figure S3: MR-proADM levels according to the occurrence of bloodstream infection or pneumonia, on days 1, 3, and 7; Table S1: Landmark analysis of the association of day-3 and day-7 MR-proADM with day-60 mortality, according to day-1 MR-proADM.

Author Contributions: Conceptualization, E.d.M., K.P., S.R., L.B., R.S. and J.-F.T.; methodology, E.d.M., K.P., S.R. and J.-F.T.; formal analysis, S.R.; investigation, E.d.M., M.M., P.-H.W., J.P. and P.J.; resources, K.P. and J.-F.T.; data curation, S.R.; writing—original draft preparation, E.d.M.; writing—review and editing, E.d.M., K.P., M.M., S.R., P.-H.W., J.P., P.J., R.S., L.B. and J.-F.T.; visualization, E.d.M. and S.R.; supervision, J.-F.T.; project administration, S.R. and J.-F.T.; funding acquisition, K.P. and J.-F.T. All authors have read and agreed to the published version of the manuscript.

Funding: This research was funded by a clinical research agreement from THERMO FISHER SCIENTIFIC (BRAHMS GmbH, Hennigsdorf, Germany).

Institutional Review Board Statement: The study was conducted in accordance with the Declaration of Helsinki and approved by the Institutional Review Board (or Ethics Committee) of the French society of intensive care (CE-SRLF 22-76); 12 September 2021.

Informed Consent Statement: Informed consent was obtained from all subjects involved in the study.

Data Availability Statement: The datasets used and analyzed during the current study are available from the corresponding author on reasonable request.

Conflicts of Interest: The authors received a research grant from Thermo Fisher to perform the study. Outside of the study, J.-F.T. declares a participation to the boards of Merck, Pfizer, BD, Gilead and Paratek, and lecture fees for Pfizer, Shionogi, BD and Merck.

References

1. WHO. Coronavirus (COVID-19) Dashboard. Available online: https://covid19.who.int (accessed on 22 August 2022).
2. COVID-ICU Group on behalf of the REVA Network and the COVID-ICU Investigators. Clinical Characteristics and Day-90 Outcomes of 4244 Critically Ill Adults with COVID-19: A Prospective Cohort Study. *Intensive Care Med.* **2021**, *47*, 60–73. [CrossRef]
3. Varga, Z.; Flammer, A.J.; Steiger, P.; Haberecker, M.; Andermatt, R.; Zinkernagel, A.S.; Mehra, M.R.; Schuepbach, R.A.; Ruschitzka, F.; Moch, H. Endothelial Cell Infection and Endotheliitis in COVID-19. *Lancet* **2020**, *395*, 1417–1418. [CrossRef]
4. Bonaventura, A.; Vecchié, A.; Dagna, L.; Martinod, K.; Dixon, D.L.; Van Tassell, B.W.; Dentali, F.; Montecucco, F.; Massberg, S.; Levi, M.; et al. Endothelial Dysfunction and Immunothrombosis as Key Pathogenic Mechanisms in COVID-19. *Nat. Rev. Immunol.* **2021**, *21*, 319–329. [CrossRef] [PubMed]
5. Calabretta, E.; Moraleda, J.M.; Iacobelli, M.; Jara, R.; Vlodavsky, I.; O'Gorman, P.; Pagliuca, A.; Mo, C.; Baron, R.M.; Aghemo, A.; et al. COVID-19-induced Endotheliitis: Emerging Evidence and Possible Therapeutic Strategies. *Br. J. Haematol.* **2021**, *193*, 43–51. [CrossRef] [PubMed]

6. Temmesfeld-Wollbrück, B.; Brell, B.; Dávid, I.; Dorenberg, M.; Adolphs, J.; Schmeck, B.; Suttorp, N.; Hippenstiel, S. Adrenomedullin Reduces Vascular Hyperpermeability and Improves Survival in Rat Septic Shock. *Intensive Care Med.* **2007**, *33*, 703–710. [CrossRef]
7. Christ-Crain, M.; Morgenthaler, N.G.; Struck, J.; Harbarth, S.; Bergmann, A.; Müller, B. Mid-Regional pro-Adrenomedullin as a Prognostic Marker in Sepsis: An Observational Study. *Crit. Care* **2005**, *9*, R816. [CrossRef] [PubMed]
8. Elke, G.; Bloos, F.; Wilson, D.C.; Brunkhorst, F.M.; Briegel, J.; Reinhart, K.; Loeffler, M.; Kluge, S.; Nierhaus, A.; Jaschinski, U.; et al. The Use of Mid-Regional Proadrenomedullin to Identify Disease Severity and Treatment Response to Sepsis-a Secondary Analysis of a Large Randomised Controlled Trial. *Crit. Care* **2018**, *22*, 79. [CrossRef]
9. Ara-Somohano, C.; Bonadona, A.; Carpentier, F.; Pavese, P.; Vesin, A.; Hamidfar-Roy, R.; Minet, C.; Vanzetto, G.; Schwebel, C.; Timsit, J.-F. Evaluation of Eight Biomarkers to Predict Short-Term Mortality in Patients with Acute Severe Dyspnea. *Minerva Anestesiol.* **2017**, *83*, 824–835. [CrossRef]
10. Liu, D.; Xie, L.; Zhao, H.; Liu, X.; Cao, J. Prognostic Value of Mid-Regional pro-Adrenomedullin (MR-ProADM) in Patients with Community-Acquired Pneumonia: A Systematic Review and Meta-Analysis. *BMC Infect. Dis.* **2016**, *16*, 232. [CrossRef]
11. Li, P.; Wang, C.; Pang, S. The Diagnostic Accuracy of Mid-Regional pro-Adrenomedullin for Sepsis: A Systematic Review and Meta-Analysis. *Minerva Anestesiol.* **2021**, *87*, 1117–1127. [CrossRef]
12. Moore, N.; Williams, R.; Mori, M.; Bertolusso, B.; Vernet, G.; Lynch, J.; Philipson, P.; Ledgerwood, T.; Kidd, S.P.; Thomas, C.; et al. Mid-Regional Proadrenomedullin (MR-ProADM), C-Reactive Protein (CRP) and Other Biomarkers in the Early Identification of Disease Progression in Patients with COVID-19 in the Acute NHS Setting. *J. Clin. Pathol.* **2022**. *Epub ahead of print*. [CrossRef] [PubMed]
13. Zaninotto, M.; Mion, M.M.; Marchioro, L.; Padoan, A.; Plebani, M. Endothelial Dysfunction and Mid-Regional ProAdrenomedullin: What Role in SARS-CoV-2 Infected Patients? *Clin. Chim. Acta* **2021**, *523*, 185–190. [CrossRef] [PubMed]
14. Gregoriano, C.; Koch, D.; Kutz, A.; Haubitz, S.; Conen, A.; Bernasconi, L.; Hammerer-Lercher, A.; Saeed, K.; Mueller, B.; Schuetz, P. The Vasoactive Peptide MR-pro-Adrenomedullin in COVID-19 Patients: An Observational Study. *Clin. Chem. Lab. Med.* **2021**, *59*, 995–1004. [CrossRef]
15. Sozio, E.; Tascini, C.; Fabris, M.; D'Aurizio, F.; De Carlo, C.; Graziano, E.; Bassi, F.; Sbrana, F.; Ripoli, A.; Pagotto, A.; et al. MR-ProADM as Prognostic Factor of Outcome in COVID-19 Patients. *Sci. Rep.* **2021**, *11*, 5121. [CrossRef] [PubMed]
16. Minieri, M.; Di Lecce, V.N.; Lia, M.S.; Maurici, M.; Bernardini, S.; Legramante, J.M. Role of MR-ProADM in the Risk Stratification of COVID-19 Patients Assessed at the Triage of the Emergency Department. *Crit. Care Lond. Engl.* **2021**, *25*, 407. [CrossRef]
17. De Guadiana-Romualdo, L.G.; Martínez, M.M.; Mulero, M.D.R.; Esteban-Torrella, P.; Olivo, M.H.; García, M.J.A.; Campos-Rodríguez, V.; Sancho-Rodríguez, N.; Martínez, M.G.; Alcaraz, A.; et al. Circulating MR-ProADM Levels, as an Indicator of Endothelial Dysfunction, for Early Risk Stratification of Mid-Term Mortality in COVID-19 Patients. *Int. J. Infect. Dis. IJID Off. Publ. Int. Soc. Infect. Dis.* **2021**, *111*, 211–218. [CrossRef]
18. Van Oers, J.A.H.; Kluiters, Y.; Bons, J.A.P.; de Jongh, M.; Pouwels, S.; Ramnarain, D.; de Lange, D.W.; de Grooth, H.-J.; Girbes, A.R.J. Endothelium-Associated Biomarkers Mid-Regional Proadrenomedullin and C-Terminal Proendothelin-1 Have Good Ability to Predict 28-Day Mortality in Critically Ill Patients with SARS-CoV-2 Pneumonia: A Prospective Cohort Study. *J. Crit. Care* **2021**, *66*, 173–180. [CrossRef]
19. Oblitas, C.-M.; Galeano-Valle, F.; Ramírez-Navarro, J.; López-Cano, J.; Monterrubio-Manrique, Á.; García-Gámiz, M.; Sancho-González, M.; Arenal-López, S.; Álvarez-Sala Walther, L.-A.; Demelo-Rodríguez, P. Mid-Regional Pro-Adrenomedullin, Methemoglobin and Carboxyhemoglobin as Prognosis Biomarkers in Critically Ill Patients with COVID-19: An Observational Prospective Study. *Viruses* **2021**, *13*, 2445. [CrossRef]
20. Montrucchio, G.; Sales, G.; Rumbolo, F.; Palmesino, F.; Fanelli, V.; Urbino, R.; Filippini, C.; Mengozzi, G.; Brazzi, L. Effectiveness of Mid-Regional pro-Adrenomedullin (MR-ProADM) as Prognostic Marker in COVID-19 Critically Ill Patients: An Observational Prospective Study. *PLoS ONE* **2021**, *16*, e0246771. [CrossRef]
21. Lippi, G.; Henry, B.M. Pooled Analysis of Mid-Regional pro-Adrenomedullin Values in COVID-19 Patients with Critical Illness. *Intern. Emerg. Med.* **2021**, *16*, 1723–1725. [CrossRef]
22. Musuuza, J.S.; Watson, L.; Parmasad, V.; Putman-Buehler, N.; Christensen, L.; Safdar, N. Prevalence and Outcomes of Co-Infection and Superinfection with SARS-CoV-2 and Other Pathogens: A Systematic Review and Meta-Analysis. *PLoS ONE* **2021**, *16*, e0251170. [CrossRef] [PubMed]
23. Wicky, P.-H.; Niedermann, M.S.; Timsit, J.-F. Ventilator-Associated Pneumonia in the Era of COVID-19 Pandemic: How Common and What Is the Impact? *Crit. Care Lond. Engl.* **2021**, *25*, 153. [CrossRef] [PubMed]
24. Segala, F.V.; Bavaro, D.F.; Di Gennaro, F.; Salvati, F.; Marotta, C.; Saracino, A.; Murri, R.; Fantoni, M. Impact of SARS-CoV-2 Epidemic on Antimicrobial Resistance: A Literature Review. *Viruses* **2021**, *13*, 2110. [CrossRef] [PubMed]
25. Altman, D.G.; Lausen, B.; Sauerbrei, W.; Schumacher, M. Dangers of Using "Optimal" Cutpoints in the Evaluation of Prognostic Factors. *J. Natl. Cancer Inst.* **1994**, *86*, 829–835. [CrossRef]
26. Spoto, S.; Cella, E.; de Cesaris, M.; Locorriere, L.; Mazzaroppi, S.; Nobile, E.; Lanotte, A.M.; Pedicino, L.; Fogolari, M.; Costantino, S.; et al. Procalcitonin and MR-Proadrenomedullin Combination with SOFA and QSOFA Scores for Sepsis Diagnosis and Prognosis: A Diagnostic Algorithm. *Shock* **2018**, *50*, 44–52. [CrossRef]

27. Charles, P.-E.; Péju, E.; Dantec, A.; Bruyère, R.; Meunier-Beillard, N.; Dargent, A.; Prin, S.; Wilson, D.; Quenot, J.-P. Mr-Proadm Elevation Upon Icu Admission Predicts the Outcome of Septic Patients and Is Correlated with Upcoming Fluid Overload. *Shock* **2017**, *48*, 418–426. [CrossRef]
28. Stalenhoef, J.E.; van Nieuwkoop, C.; Wilson, D.C.; van der Starre, W.E.; Delfos, N.M.; Leyten, E.M.S.; Koster, T.; Ablij, H.C.; Van't Wout, J.J.W.; van Dissel, J.T. Biomarker Guided Triage Can Reduce Hospitalization Rate in Community Acquired Febrile Urinary Tract Infection. *J. Infect.* **2018**, *77*, 18–24. [CrossRef]
29. Oussalah, A.; Lagneaux, A.-S.; Alix, T.; Filhine-Tresarrieu, P.; Callet, J.; Ferrand, J.; Jung, J.; Broseus, J.; Salignac, S.; Luc, A.; et al. Ascitic Fluid Mid-Regional-pro-Adrenomedullin (MR-pro-ADM): A Novel Rapid-Assay Sepsis Biomarker to Diagnose Spontaneous Bacterial Peritonitis in Cirrhotic Patients. *J. Hepatol.* **2022**, *77*, S905–S906. [CrossRef]
30. Pickens, C.O.; Gao, C.A.; Cuttica, M.J.; Smith, S.B.; Pesce, L.L.; Grant, R.A.; Kang, M.; Morales-Nebreda, L.; Bavishi, A.A.; Arnold, J.M.; et al. Bacterial Superinfection Pneumonia in Patients Mechanically Ventilated for COVID-19 Pneumonia. *Am. J. Respir. Crit. Care Med.* **2021**, *204*, 921–932. [CrossRef]
31. Bhakta, S.; Sanghavi, D.K.; Johnson, P.W.; Kunze, K.L.; Neville, M.R.; Wadei, H.M.; Bosch, W.; Carter, R.E.; Shah, S.Z.; Pollock, B.D.; et al. Clinical and Laboratory Profiles of the SARS-CoV-2 Delta Variant Compared with Pre-Delta Variants. *Int. J. Infect. Dis. IJID Off. Publ. Int. Soc. Infect. Dis.* **2022**, *120*, 88–95. [CrossRef]
32. Bouzid, D.; Visseaux, B.; Kassasseya, C.; Daoud, A.; Fémy, F.; Hermand, C.; Truchot, J.; Beaune, S.; Javaud, N.; Peyrony, O.; et al. Comparison of Patients Infected with Delta Versus Omicron COVID-19 Variants Presenting to Paris Emergency Departments: A Retrospective Cohort Study. *Ann. Intern. Med.* **2022**, *175*, 831–837. [CrossRef] [PubMed]
33. Le Gall, J.R.; Lemeshow, S.; Saulnier, F. A New Simplified Acute Physiology Score (SAPS II) Based on a European/North American Multicenter Study. *JAMA* **1993**, *270*, 2957–2963. [CrossRef] [PubMed]
34. Vincent, J.L.; Moreno, R.; Takala, J.; Willatts, S.; De Mendonça, A.; Bruining, H.; Reinhart, C.K.; Suter, P.M.; Thijs, L.G. The SOFA (Sepsis-Related Organ Failure Assessment) Score to Describe Organ Dysfunction/Failure. On Behalf of the Working Group on Sepsis-Related Problems of the European Society of Intensive Care Medicine. *Intensive Care Med.* **1996**, *22*, 707–710. [CrossRef] [PubMed]
35. Vesin, A.; Azoulay, E.; Ruckly, S.; Vignoud, L.; Rusinovà, K.; Benoit, D.; Soares, M.; Azeivedo-Maia, P.; Abroug, F.; Benbenishty, J.; et al. Reporting and Handling Missing Values in Clinical Studies in Intensive Care Units. *Intensive Care Med.* **2013**, *39*, 1396–1404. [CrossRef] [PubMed]

antibiotics 2022, *11*, 1141

Article

Diagnostic Accuracy of Procalcitonin upon Emergency Department Admission during SARS-CoV-2 Pandemic

Stefano Malinverni [1,*], Silvia Lazzaroni [1], Maïa Nuñez [2], Thierry Preseau [2], Frédéric Cotton [3], Delphine Martiny [4], Fatima Bouazza [1], Vincent Collot [1], Deborah Konopnicki [5], Stéphane Alard [6] and Magali Bartiaux [1]

1 Emergency Department, Centre Hospitalier Universitaire Saint-Pierre, Université Libre de Bruxelles, Rue Haute 322, 1000 Brussels, Belgium
2 Centre Hospitalier Universitaire Brugmann, Place A.Van Gehuchten 4, Université Libre de Bruxelles, 1020 Brussels, Belgium
3 Clinical Chemistry, Laboratoire Hospitalier Universitaire de Bruxelles-Universitair Laboratorium Brussel (LHUB-ULB), Université Libre de Bruxelles, Rue Haute 322, 1000 Brussels, Belgium
4 Department of Microbiology, Laboratoire Hospitalier Universitaire de Bruxelles-Universitair Laboratorium Brussel (LHUB-ULB), Rue Haute 322, 1000 Brussels, Belgium
5 Infectious Diseases Department, Centre Hospitalier Universitaire Saint-Pierre, Université Libre de Bruxelles, Rue Haute 322, 1000 Brussels, Belgium
6 Department of Radiology, Centre Hospitalier Universitaire Saint-Pierre, Université Libre de Bruxelles, Rue Haute 322, 1000 Brussels, Belgium
* Correspondence: stefano.malinverni@stpierre-bru.be

Citation: Malinverni, S.; Lazzaroni, S.; Nuñez, M.; Preseau, T.; Cotton, F.; Martiny, D.; Bouazza, F.; Collot, V.; Konopnicki, D.; Alard, S.; et al. Diagnostic Accuracy of Procalcitonin upon Emergency Department Admission during SARS-CoV-2 Pandemic. *Antibiotics* **2022**, *11*, 1141. https://doi.org/10.3390/antibiotics11091141

Academic Editors: Sabine Danielle Allard and Johan Van Laethem

Received: 7 July 2022
Accepted: 18 August 2022
Published: 23 August 2022

Highlights:

- Procalcitonin has low sensitivity for bacterial pneumonia at emergency admission.
- Procalcitonin has low specificity for bacterial pneumonia at emergency admission.
- Procalcitonin sampled at emergency admission should not guide antibiotic prescriptions.
- Higher procalcitonin values in COVID-19 pneumonia hinder its use for antibiotic stewardship.
- No procalcitonin cutoff level provided reliable guidance in antibiotic prescription.

Abstract: Introduction: Procalcitonin is a marker for bacterial diseases and has been used to guide antibiotic prescription. Procalcitonin accuracy, measured at admission, in patients with community-acquired pneumonia (CAP), is unknown in the current severe acute respiratory syndrome coronavirus 2 (SARS-CoV-2) pandemic. Objectives: To evaluate the diagnostic accuracy of procalcitonin to assess the need for antibiotic treatment in patients with CAP presenting to the emergency department during the SARS-CoV-2 pandemic. Methods: We performed a real-world diagnostic retrospective accuracy study of procalcitonin in patients admitted to the emergency department. Measures of diagnostic accuracy were calculated based on procalcitonin results compared to the reference standard of combined microbiological and radiological analysis. Sensitivity, specificity, positive and negative predictive values, and area under (AUC) the receiver-operating characteristic (ROC) curve were calculated in two analyses: first assessing procalcitonin ability to differentiate microbiologically proven bacteria from viral CAP and then clinically diagnosed bacterial CAP from viral CAP. Results: When using a procalcitonin threshold of 0.5 ng/mL to identify bacterial etiology within patients with CAP, we observed sensitivity and specificity of 50% and 64.1%, and 43% and 82.6%, respectively, in the two analyses. The positive and negative predictive values of a procalcitonin threshold of 0.5 ng/mL to identify patients for whom antibiotics should be advised were 46.4% and 79.7%, and 48.9% and 79% in the two analyses, respectively. The AUC for the two analyses was 0.60 (95% confidence interval [CI] 0.52–0.68) and 0.62 (95% CI, 0.55–0.69). Conclusions: Procalcitonin measured upon admission during the SARS-CoV-2 pandemic should not guide antibiotic treatment in patients with CAP.

Keywords: COVID-19; pandemics; procalcitonin; SARS virus; community-acquired infections; emergency service; hospital; pneumonia; viral; community-acquired pneumonia

1. Introduction

Severe acute respiratory syndrome coronavirus 2 (SARS-CoV-2) infection is associated with high rates of emergency attendance, hospitalization, and intensive care unit admission. Most cases present with mild symptoms, and a small proportion evolve to more severe presentations, such as oxygen-requiring pneumonia, acute respiratory distress syndrome, or fatal issues [1]. The most common symptoms are fever, fatigue, and dry cough. Less common symptoms include sputum production, anorexia, sore throat, chest pain, and nausea [1,2]. These symptoms are aspecific and are frequently observed in pneumonia caused by other viruses and bacteria. Differentiating between viral and bacterial pneumonia or bacterial coinfection of viral pneumonia is challenging because of the overlap in presentation between these entities [3,4]. The choice of administering broad-spectrum antibiotics to these patients is difficult, and overprescription of antibiotics has been reported in hospitalized patients with SARS-CoV-2 [5]. While a delay in antibiotic treatment of bacterial community-acquired pneumonia (CAP) is associated with increased mortality [6], systematic broad-spectrum antibiotic treatment of suspected bacterial CAP is associated with complications, side effects, and mortality [7].

Procalcitonin (PCT) is a prohormone produced by the thyroid gland. In response to bacterial infection, interleukin-6 (IL-6), tumor necrosis factor-α (TNF-α), and interleukin-1β (IL-β) induce PCT synthesis in extrathyroidal tissue [8] with a peak at 6 h from the onset of infection and a half-life of 24 h [9]. In most viral infections, increased interferon gamma production inhibits PCT synthesis, leading to bacterial specificity of PCT.

Studies have suggested that PCT is a useful serum biomarker that supports clinical decisions regarding antibiotic treatment in patients with CAP. Higher serum PCT levels are associated with a higher probability of bacterial disease [10]. Some clinical trials have reported that PCT can guide clinicians in decisions on empiric antibiotic coverage without incurring higher rates of adverse outcomes [11,12], while others have reported no differences in antibiotic use among patients with suspected lower respiratory infection when PCT values were integrated in the treatment decision [4].

Within the context of the ongoing pandemic, PCT has been associated with the severity and mortality of SARS-CoV-2 infections [13–15]. Meanwhile, evidence on the role of PCT in guiding antibiotic prescriptions remains insufficient [16,17].

This two-center case–control study aimed to evaluate the role of PCT in differentiating CAP with an indication for antibiotic treatment from other entities associated with a new radiological infiltrate and lower respiratory tract infection (LRTI) signs and symptoms.

2. Material and Methods

2.1. Design

This retrospective two-center case–control observational study assessed the diagnostic accuracy of PCT for antibiotic prescription guidance in patients with CAP admitted to the Emergency Department of Saint-Pierre and Brugmann University hospitals between 1 March 2020 and 31 October 2020. The ethics committee of each hospital (OM 007 and OM 026) approved the study protocol and waived the need for signed informed consent due to the retrospective design of the study (CE20-12-11 and CE 2022/132).

2.2. Setting and Participants

A sample size of 225 was calculated as sufficient to detect specificity and sensitivity of 0.8 with a two-sided type I error of 0.05 and a power of 80%, assuming a prevalence of 20% and H_0 of 0.6 for both sensitivity and specificity. For eligibility, we assessed a convenience sample of consecutive patient consultations and expected to achieve the previously calculated sample size within the study timeframe. Both centers routinely measure PCT levels of patients with suspected CAP upon admission. Patients with CAP having a serum PCT measurement performed within 24 h from ED admission and at least one viral and one bacterial investigation (a pair of hemocultures, sputum, or bronchoalveolar lavage) performed within 48 h from admission were included in this study. Minors, pregnant women, patients already on antibiotic treatment at the time of ED admission, and patients with an extrapulmonary site of infection diagnosed during their initial evaluation were excluded.

All enrolled patients had signs of acute infection (temperature of >38 °C, chills, altered mental status, and leukocyte count of >10,000/µL or <4000/µL). Moreover, all patients had at least one symptom of acute respiratory illness (cough, dyspnea, sputum production, tachypnea, pleuritic chest pain, ambient air oxygen saturation ($SatO_2$) of <94%, or a loss of $\geq$4 $SatO_2$ points following a 1 min walking test) or at least one finding during auscultation (crackles and rales). All included patients had a new infiltrate on radiological imaging performed within 48 h from admission. Missing data were treated as missing in the analysis, and no imputation was performed.

2.3. Outcome Measure and Analysis

PCT concentrations were measured by technicians blinded to the clinical information using Lumipulse G B•R•A•H•M•S PCT immunoreaction cartridges on a Lumipulse G600II instrument (Fujirebio, Ghent, Belgium).

Bacteriological analysis included culture from a respiratory tract specimen (sputum or bronchoalveolar lavage) or a pair of blood cultures. Diagnosis of SARS-CoV-2 infection was based on the COVID-19 Ag Respi-Strip® (Coris Bioconcept, Gembloux, Belgium), followed by qRT-PCR in case of a negative result (RealStar® SARS-CoV-2 RT PCR Kit, Altona Diagnostics), both performed on a nasopharyngeal swab [18]. Additional microbiological tests were performed according to clinical presentation. Additional bacterial testing included the *Legionella pneumophila* urinary antigen test and serological tests for *Chlamydophila pneumoniae* and *Mycoplasma pneumonia*. Additional viral testing included immunochromatographic techniques for influenza and adenovirus, direct fluorescence antibody tests for parainfluenza viruses, and inoculation of three cell cultures. Upon request, for 13 patients, respiratory tract specimens were analyzed using a multiplex PCR system to detect an additional fourteen viral and three bacterial targets (Biofire™ Filmarray™, bioMérieux, Marcy l'Etoile, France).

Chest radiographs were classified as showing radiographic evidence of pneumonia whenever a new infiltrate was mentioned in the X-ray report. Thoracic CT scans were considered to have radiographic evidence of pneumonia when the summary report mentioned probable bacterial or viral pneumonia. Radiological assessment was performed by a radiologist blinded to the PCT results. During the study period, thoracic low-dose CT scans for suspected CAP were additionally classified through a simplified classification as having typical features of bacterial CAP, typical features of SARS-CoV-2 CAP, intermediate features of both viral and bacterial CAP, or features evoking a diagnosis different from CAP.

CAP was defined as a new infiltrate on chest radiological study in a patient presenting with LRTI signs and symptoms [19]. Patients were classified according to laboratory test results and interpretation of radiological images. CAP for which antibiotic treatment was recommended was defined as cases with a microbiological analysis positive for pathogenic bacteria or with a chest CT lung infiltrate typical of bacterial pneumonia. CAP for which antibiotic treatment was recommended included bacterial CAP and viral CAP cases with a documented bacterial coinfection. CAP cases in which antibiotics were discouraged were defined as cases with a typical viral pneumonia infiltrate on radiological studies, with a microbiological analysis positive for a pathogenic virus and a lack of any microbiological analysis positive for a bacterial pathogen or imaging test suggesting a possible bacterial coinfection.

Sixty-eight (19%) patients could not be classified using this method. These cases were classified, in addition to the previously categorized patients, by two independent specialists blinded to the PCT results in a secondary analysis. These two specialists classified ambiguous cases according to clinical, microbiological, and radiographic results, antibiotic administration, and clinical evolution as CAP cases for which antibiotic treatment was recommended. In cases of disagreement, a third independent specialist provided a definitive classification.

Therefore, we retrospectively studied the accuracy of PCT in identifying CAP cases in which antibiotic therapy was recommended in two separate analyses, using the two aforementioned definitions for recommended or discouraged antibiotic treatment.

3. Analysis

3.1. Procalcitonin among Groups

We compared the PCT distribution within the two aforementioned analyses using the Wilcoxon rank-sum test. PCT cutoff values described in the literature as thresholds for identifying bacterial infections and guiding antibiotic therapy [11,12] were used to categorize CAP according to PCT values in four strata: <0.1 ng/mL, 0.1–0.249 ng/mL, 0.25–0.499 ng/mL, and >0.5 ng/mL.

3.2. Accuracy of Procalcitonin for Identifying Antibiotic-Requiring CAP

We calculated a nonparametric receiver-operating characteristic (ROC) curve for the two aforementioned dichotomous analyses to study the diagnostic accuracy of PCT in identifying antibiotic-requiring CAP. Sensitivity, specificity, negative predictive values, and positive predictive values were calculated using PCT cutoff values of 0.1 ng/mL, 0.25 ng/mL and 0.5 ng/mL. All statistical analyses were performed using Stata software version 16 (StataCorp, College Station, TX, USA).

4. Results

During the study period, 476 patients presented with complaints related to CAP. After application of inclusion and exclusion criteria, 359 (75.4%) patients were included in the current analysis (Figure 1).

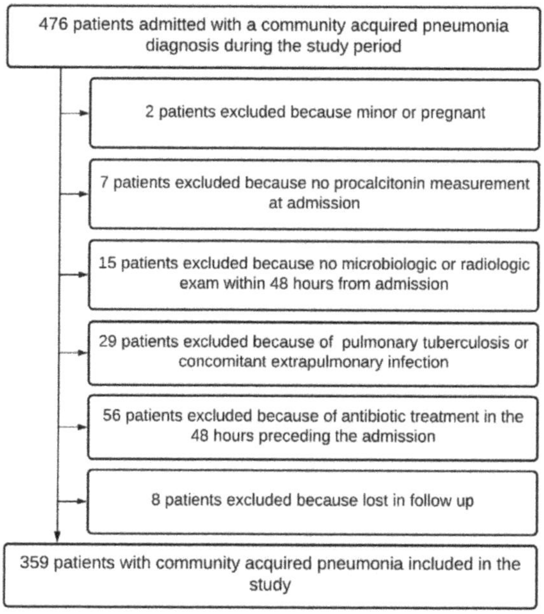

Figure 1. Study-inclusion flowchart.

Patient characteristics at inclusion are illustrated in Table 1.

Table 1. Demographic and clinical characteristics at study inclusion.

Number of Observations	359
Age, mean (IQR), y	61 (49–75)
Female, No. (%)	112 (31.2)
Days since first symptoms, median (IQR)	7 (4–9)
Coexisting conditions, No. (%)	
Hypertension	148 (41.2)
Diabetes	105 (29.3)
Chronic renal failure	59 (16.6)
COPD	27 (7.6)
Asthma	24 (6.7)
Chronic heart failure	21 (5.9)
Cerebrovascular disease	17 (4.7)
Nursing home resident, No. %	40 (11.1)
Pneumonia severity index, median (IQR)	77 (58–105)
Heart rate, median (IQR), beats/min	99 (85–111)
PaO_2/FiO_2, median (IQR)	300 (245–340)
Respiratory rate, median, (IQR), breaths/min	24 (20–30)
Systolic blood pressure, mean (IQR), mmHg	130 (117–144)
Blood urea nitrogen, median (IQR), mg/dL	16 (11–22)
Arterial pH, median (IQR)	7.47 (7.45–7.50)
Lactate, median (IQR), mEq/L	1.1 (0.9–1.7)
Procalcitonin, median (IQR) ng/mL	0.19 (0.10–0.48)

Abbreviations: No., number; IQR, interquartile range; COPD, chronic obstructive pulmonary disease; PaO_2/FiO_2, arterial oxygen partial pressure to fractional inspired oxygen ratio.

Table 2 illustrates the etiological pathogens identified.

Table 2. Etiological diagnosis using microbiological and molecular methods.

Number of Observations	359
Identified pathogen, No. (%)	297 (83)
SARS-CoV-2	244 (68)
Staphylococcus aureus	18 (5)
Hemophylus influenzae	9 (2.5)
Streptococcus pneumoniae	8 (2.2)
Other	28 (7.8)

Abbreviations: No., number.

In total, 77 cases (26.5%) were classified as having CAP with an indication for antibiotic treatment in the primary analysis and 100 (27.9%) in the secondary analysis. All 214 (100%) patients with microbiologically documented viral CAP had SARS-CoV-2-related pneumonia.

PCT concentrations were higher in the CAP group where antibiotics were recommended (0.22 ng/mL; IQR, 0.11–2.22 ng/mL) than the CAP group for which antibiotic were discouraged (0.19 ng/mL; IQR, 0.1–0.39 ng/mL; $p = 0.01$) (Supplementary Digital Content; Figure S1).

Antibiotic-requiring CAP was more prevalent in higher PCT strata. The prevalence of antibiotic-requiring CAP was 24% among patients with PCT <0.1 ng/mL and increased to 46% among patients with PCT $\geq$ 0.5 ng/mL (Supplementary Digital Content Table S1). Results were ambiguous in the intermediate strata.

4.1. *Accuracy of PCT for Identifying CAP Cases in Which Antibiotic Therapy Was Recommended*

4.1.1. Nested Cohort without Cases Classified According to Specialist Opinion

PCT performed poorly in identifying cases for which antibiotics were recommended (area under the curve [AUC], 0.60; 95% confidence interval [CI], 0.52–0.68) in the nested cohort, excluding patients classified according to specialist opinion (Figure 2).

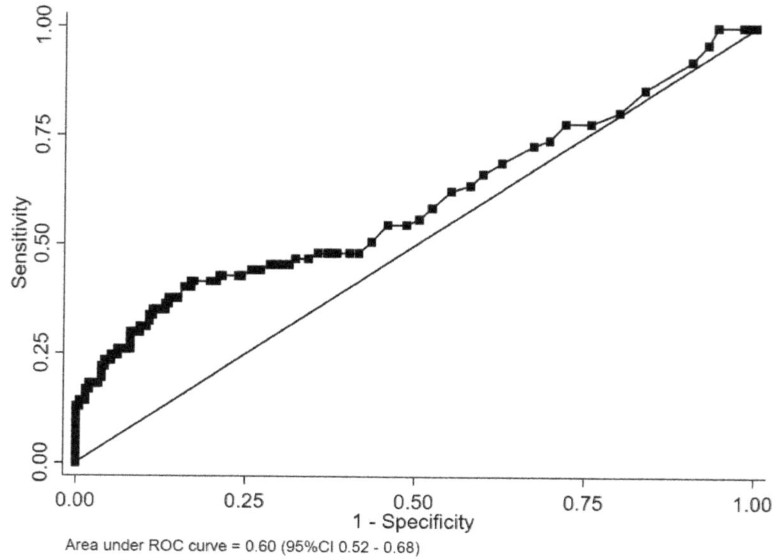

Figure 2. Receiver-operating characteristic curve of serum procalcitonin for the diagnosis of microbiologically proven bacterial pneumonia.

A PCT threshold of ≥0.25 ng/mL to identify CAP for which antibiotics were recommended resulted in sensitivity of 48.1% (95% CI, 36.5–59.7%) and specificity of 61.7% (95% CI, 54.8–68.2%) (Table 3).

Table 3. Cutoff levels, sensitivity and specificity for procalcitonin in detecting bacterial pneumonia.

Value	Cutoff Level ng/mL	Sensitivity, % (95% CI)	Specificity, % (95% CI)	Patients, No. (%)				PPV	NPV
				True Positive	False Negative	False Positive	True Negative	(95% CI)	(95% CI)
Procalcitonin [antibiotic-requiring vs. viral CAP] n = 291	>0.1	77.9 (67–86.6)	24.3 (18.7–30.6)	60 (20.6)	17 (5.9)	162 (55.7)	52 (17.9)	27.0 (21.3–33.4)	75.4 (63.5–84.9)
	>0.25	48.1 (36.5–59.7)	61.7 (54.8–68.2)	37 (12.7)	40 (13.8)	82 (28.2)	132 (45.4)	31.1 (22.9–40.2)	76.7 (69.7–82.8)
	≥0.5	41.6 (30.4–53.4)	82.7 (77.0–87.5)	32 (11.0)	45 (15.5)	37 (12.7)	177 (60.8)	46.4 (34.3–58.8)	79.7 (73.8–84.8)
Procalcitonin [confirmed and clinical bacterial vs. confirmed and clinical viral CAP] n = 359	>0.1	79.0 (69.7–86.5)	25.5 (20.3–31.2)	79 (22.0)	21 (5.9)	193 (53.8)	66 (18.3)	29.0 (23.7–34.8)	75.9 (65.5–84.4)
	>0.25	50.0 (39.8–60.2)	64.1 (57.9–69.9)	50 (13.9)	50 (13.9)	93 (25.9)	166 (46.2)	35.0 (27.2–43.4)	76.9 (70.6–82.3)
	≥0.5	43.0 (33.1–53.3)	82.6 (77.5–87.0)	43 (12.0)	57 (15.9)	45 (12.5)	214 (59.6)	48.9 (38.1–59.8)	79.0 (73.6–83.7)

Abbreviations: PPV, positive predictive value; NPV, negative predictive value.

A threshold of ≥0.5 ng/mL to identify CAP in which antibiotics were recommended resulted in sensitivity of 41.6% (95% CI: 30.4–53.4%) and specificity of 82.7% (95% CI: 77.0–87.5%).

4.1.2. Complete Cohort including Patients Classified According to Specialist Opinion

PCT performed poorly in identifying CAP cases for whom antibiotics were recommended [AUC: 0.62 (95% CI: 0.55–0.69)] in the entire cohort (Figure 3).

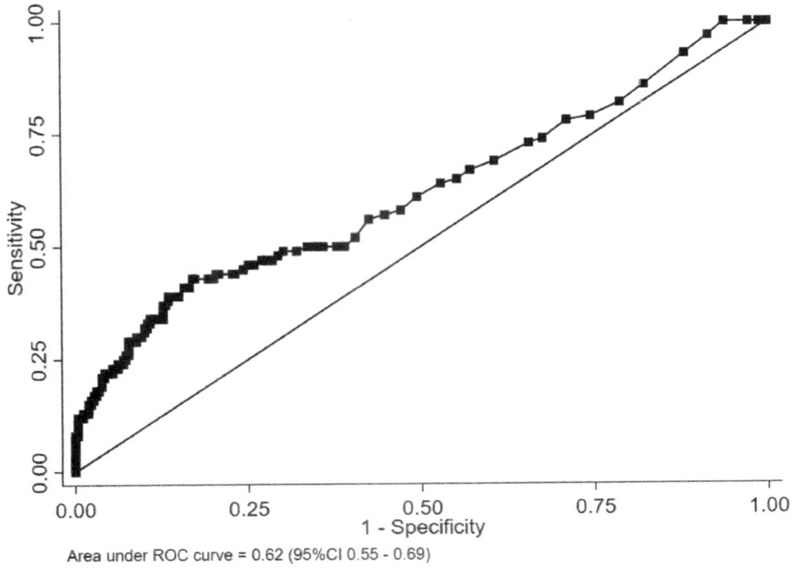

Figure 3. Receiver-operating characteristic curve of serum procalcitonin for the diagnosis of clinical bacterial pneumonia.

A PCT threshold of ≥0.25 ng/mL to identify CAP for which treatment with antibiotics was indicated resulted in a sensitivity of 50.0% (95% CI, 39.8–60.2%) and a specificity of 64.1% (95% CI, 57.9–69.9%) (Table 2). A PCT threshold of ≥0.5 ng/mL to identify CAP for which treatment with antibiotic was indicated resulted in sensitivity of 43% (95% CI, 33.1–53.3%) and specificity of 82.6% (95% CI, 77.5–87.0%).

We performed a sensitivity analysis nested on patients with SARS-CoV-2 to study the accuracy of procalcitonin in identifying a bacterial coinfection within CAP patients with SARS-CoV-2. Results were consistent with previous analysis. A threshold of ≥0.5 ng/mL to identify bacterial coinfections within SARS-CoV-2 CAP resulted in sensitivity of 40% (95% CI: 16.3–67.7%) and specificity of 80.4% (95% CI: 74.9–85.1%). PCT performed poorly in identifying bacterial coinfection (area under the curve [AUC], 0.60; 95% confidence interval [CI], 0.52–0.68) within SARS-CoV-2 CAP patients [AUC: 0.59 (95% CI: 0.44–0.76)].

5. Discussion

In this multicenter retrospective study of 359 adults with CAP admitted to the ED during the SARS-CoV-2 pandemic, including 298 with microbiologically documented pathogens, no PCT threshold identified CAP for which antibiotic treatment was recommended.

Trials and meta-analyses, partially performed in EDs, suggested that PCT could tailor antibiotic prescription in CAP without increasing adverse outcomes [11,12,20]. Guidelines based on previous results of trials provided graded recommendations based on four tiers of PCT levels and discouraged the use of antibiotics for patients with PCT values ≤0.1 ng/mL while strongly recommending antibiotics in patients with PCT values ≥0.5 ng/mL. Other studies have suggested thresholds of 0.2 ng/mL to differentiate CAP from bronchitis [21]. Studies performed during the H1N1 pandemic indicated that PCT could help distinguish patients with a bacterial etiology from those with viral pneumonia [22,23]. However, a recent trial cast doubt on the ability of PCT to reduce antibiotic exposure in ED-diagnosed CAP [4]. The results from our study, carried out during the SARS-CoV-2 pandemic, further challenged PCT-based recommendations to guide antibiotic administration in CAP. In this cohort, withholding antibiotic treatment in patients with PCT levels ≤0.1 ng/mL would have resulted in undertreating 17 (24.6%) cases of all CAP with a microbiological indication for antibiotic treatment. Moreover, the routine administration of antibiotics in our cohort in patients with PCT ≥0.5 ng/mL would have resulted in the inappropriate overtreatment of 37 (53.6%) of 69 patients (Supplement Table S1). These high rates of over- and undertreatment, as well as the low AUC observed for PCT as a tool to identify antibiotic-requiring CAP from viral CAP, do not seem to support the role of PCT in guiding antibiotic prescription in CAP within the context of the SARS-CoV-2 pandemic.

Our results are in line with previous studies [4,10,24] and with recent American guidelines for the management of CAP that recommend against routine PCT measurements to determine the need for initial antibacterial therapy [25,26]. The first ROC analysis, restricted to patients for whom a causative microbiological pathogen was documented, reported an AUC of 0.60 without any clear cutoff point that could identify patients for whom antibiotic therapy should be recommended. PCT performed poorly in identifying cases for which antibiotics were recommended, suggesting that PCT alone is not sufficient in CAP to guide antibiotic prescription in the ED. While being methodologically sound to restrict the analysis of patients with a proven microbiological etiology or a typical focal, bacterial condensation on CT scans, results from this population are not directly generalizable to clinical practice, as pathogens may be detected in less than 40% of CAP cases. Moreover, thoracic CT is not routinely performed for the diagnosis of pneumonia [27].

The secondary analysis, including both CAP with a straightforward etiologic diagnosis and pneumonia classified according to an independent blinded review, showed a similar AUC of 0.62 for the ability of PCT to identify CAP with an indication for antibiotic treatment. This analysis yielded similar results in a more pragmatic clinical scenario where the clinician has to decide whether to initiate antibiotic treatment for CAP, irrespective of whether a pathogen will be documented from the microbiological analyses.

PCT is higher in SARS-CoV-2 CAP than in CAP associated with other viral etiologies [28]. In addition, PCT had extremely low (<10%) positive predictive values for bacterial pneumonia in a multicenter study investigating empiric antibacterial therapy for suspected SARS-CoV-2 CAP [16]. The lower discriminatory performance of PCT compared to that previously reported in the literature might be explained by the hyperinflammatory status and cytokine storm caused by SARS-CoV-2, resulting in higher PCT concentrations than in other viral CAP [14], thereby lowering the discriminatory power of PCT for bacterial infections. High levels of IL-1b and TNF-α, together with high IL-6, have been reported in SARS-CoV-2 patients [29] which might increase PCT plasma levels. Alternatively, PCT trajectories rather than absolute values have been suggested as possible markers of bacterial infection without solid evidence in favor of this practice [14,30]. This finding is relevant, as PCT-driven antibiotic prescriptions in the current pandemic context might drive antibiotic overconsumption and expose patients and hospitals to potential harmful effects. Moreover, the reported findings are in line with evidence suggesting that PCT might be a marker of severity in SARS-CoV-2 pneumonia, which would hinder its ability to be used as a diagnostic tool to withdraw or withhold antibiotic prescription [13].

6. Limitations

This study has both strengths and limitations. First, the retrospective design may have introduced a selection bias, as 24.6% of the patients with CAP were not included in the analysis according to the exclusion criteria (Figure 1). Second, the low proportion of CAP cases requiring antibiotic treatment might have reduced the diagnostic performance of procalcitonin. Third, while some studies recommend serial PCT measurements to guide antibiotic prescription in critically ill patients [30], we studied PCT at the time of ED admission. Fourth, the two-center design may limit the generalization, and fifth, the emergence of new SARS-CoV-2 variants that might elicit distinct procalcitonin responses. Finally, immunosuppression, which might alter cytokine expression in response to SARS-CoV-2 infection, could influence procalcitonin response to different CAP etiologies, and was not accounted for in this analysis.

The study's main strength is its pragmatic design and the few exclusion criteria allowing us to interpret our results on PCT accuracy in identifying bacterial CAP in real-life ED conditions. The second strength is the high proportion of patients (83.0%) with a microbiologically documented etiology of CAP, which is higher than that found in previous PCT trials.

7. Conclusions

Procalcitonin measured at ED admission performed poorly as a guide for antibiotic prescription in CAP during the SARS-CoV-2 pandemic.

Supplementary Materials: The following supporting information can be downloaded at: https://www.mdpi.com/article/10.3390/antibiotics11091141/s1, Figure S1: Box plot of serum procalcitonin comparing patients with viral and antibiotic-requiring CAP; Table S1: Prevalence of CAP class by procalcitonin strata.

Author Contributions: Conceptualization, S.M., S.L., M.N., T.P., F.C., D.M., F.B., V.C., D.K., S.A. and M.B.; methodology, S.M., S.L., M.N., T.P., F.B., V.C. and M.B.; software, F.C.; validation, S.M., S.L., M.N., T.P., F.C., D.M., F.B., V.C., D.K., S.A. and M.B.; formal analysis, S.M., S.L., M.N., F.B. and V.C.; investigation, S.M., S.L., M.N., T.P., F.C., D.M., F.B., V.C., D.K., S.A. and M.B.; resources, S.M., T.P., F.C., D.M., S.A. and M.B.; data curation, S.M., S.L., M.N. and F.B.; writing—original draft preparation, S.M. and S.L.; writing—review and editing, S.M., S.L., M.N., T.P., F.C., D.M., F.B., V.C., D.K., S.A. and M.B. All authors have read and agreed to the published version of the manuscript.

Funding: This research received no external funding.

Institutional Review Board Statement: The study was conducted in accordance with the Declaration of Helsinki, and approved by the Ethics Committee) of CHU Saint Pierre and CHU Brugmann (protocol code CE20-12-11 and CE 2022/132).

Informed Consent Statement: The two Ethics Committee waived the need for signed informed consent due to the retrospective design of the study.

Data Availability Statement: The complete dataset of the study will be available in the figsharer depository with the following DOI: https://doi.org/10.6084/m9.figshare.19283672.v1.

Acknowledgments: We acknowledge Genderini, Krebs, and Gabrowska for independently classifying ambiguous CAP cases. We acknowledge IFCSM and Amis du SIAMU for their unconditional support of this research.

Conflicts of Interest: The authors declare no conflict of interest.

References

1. Guan, W.J.; Ni, Z.Y.; Hu, Y.; Liang, W.H.; Ou, C.Q.; He, J.X.; Liu, L.; Shan, H.; Lei, C.L.; Hui, D.S.C.; et al. Clinical Characteristics of Coronavirus Disease 2019 in China. *N. Engl. J. Med.* **2020**, *382*, 1708–1720. [CrossRef]
2. Hu, B.; Guo, H.; Zhou, P.; Shi, Z.-L. Characteristics of SARS-CoV-2 and COVID-19. *Nat. Rev. Microbiol.* **2021**, *19*, 141–154. [CrossRef]
3. Musher, D.M.; Roig, I.L.; Cazares, G.; Stager, C.E.; Logan, N.; Safar, H. Can an etiologic agent be identified in adults who are hospitalized for community-acquired pneumonia: Results of a one-year study. *J. Infect.* **2013**, *67*, 11–18. [CrossRef]
4. Huang, D.T.; Yealy, D.M.; Filbin, M.R.; Brown, A.M.; Chang, C.-C.H.; Doi, Y.; Donnino, M.W.; Fine, J.; Fine, M.J.; Fischer, M.A.; et al. Procalcitonin-Guided Use of Antibiotics for Lower Respiratory Tract Infection. *N. Engl. J. Med.* **2018**, *379*, 236–249. [CrossRef]
5. Rawson, T.M.; Moore, L.S.P.; Zhu, N.; Ranganathan, N.; Skolimowska, K.; Gilchrist, M.; Satta, G.; Cooke, G.; Holmes, A. Bacterial and Fungal Coinfection in Individuals With Coronavirus: A Rapid Review To Support COVID-19 Antimicrobial Prescribing. *Clin. Infect. Dis.* **2020**, *71*, 2459–2468. [CrossRef]
6. Daniel, P.; Rodrigo, C.; McKeever, T.M.; Woodhead, M.; Welham, S.; Lim, W.S. Time to first antibiotic and mortality in adults hospitalised with community-acquired pneumonia: A matched-propensity analysis. *Thorax* **2016**, *71*, 568–570. [CrossRef]
7. Webb, B.J.; Sorensen, J.; Jephson, A.; Mecham, I.; Dean, N.C. Broad-spectrum antibiotic use and poor outcomes in community-onset pneumonia: A cohort study. *Eur. Respir. J.* **2019**, *54*, 1900057. [CrossRef]
8. Assicot, M.; Bohuon, C.; Gendrel, D.; Raymond, J.; Carsin, H.; Guilbaud, J. High serum procalcitonin concentrations in patients with sepsis and infection. *Lancet* **1993**, *341*, 515–518. [CrossRef]
9. Samsudin, I.; Vasikaran, S.D. Clinical Utility and Measurement of Procalcitonin. *Clin. Biochem. Rev.* **2017**, *38*, 59–68.

10. Self, W.H.; Balk, R.A.; Grijalva, C.G.; Williams, D.J.; Zhu, Y.; Anderson, E.J.; Waterer, G.W.; Courtney, D.M.; Bramley, A.M.; Trabue, C.; et al. Procalcitonin as a Marker of Etiology in Adults Hospitalized with Community-Acquired Pneumonia. *Clin. Infect. Dis.* **2017**, *65*, 183–190. [CrossRef]
11. Christ-Crain, M.; Stolz, D.; Bingisser, R.; Müller, C.; Miedinger, D.; Huber, P.R.; Zimmerli, W.; Harbarth, S.; Tamm, M.; Müller, B. Procalcitonin guidance of antibiotic therapy in community-acquired pneumonia: A randomized trial. *Am. J. Respir. Crit. Care Med.* **2006**, *174*, 84–93. [CrossRef] [PubMed]
12. Prohosp, T.; Controlled, R.; Thomann, R.; Falconnier, C.; Wolbers, M.; Widmer, I.; Neidert, S.; Fricker, T.; Blum, C.; Schild, U.; et al. Effect of Procalcitonin-Based Guidelines vs Standard Guidelines on Antibiotic Use in Lower Respiratory Tract Infections. *JAMA-J. Am. Med. Assoc.* **2009**, *302*, 1059–1066.
13. Vanhomwegen, C.; Veliziotis, I.; Malinverni, S.; Konopnicki, D.; Dechamps, P.; Claus, M.; Roman, A.; Cotton, F.; Dauby, N. Procalcitonin accurately predicts mortality but not bacterial infection in COVID-19 patients admitted to intensive care unit. *Ir. J. Med. Sci.* **2021**, *190*, 1649–1652. [CrossRef]
14. Hu, R.; Han, C.; Pei, S.; Yin, M.; Chen, X. Procalcitonin levels in COVID-19 patients. *Int. J. Antimicrob. Agents* **2020**, *56*, 106051. [CrossRef]
15. Xu, J.; Xu, C.; Zhang, R.; Wu, M.; Pan, C.; Li, X.; Wang, Q.; Zeng, F.; Zhu, S. Associations of procalcitonin, C-reaction protein and neutrophil-to-lymphocyte ratio with mortality in hospitalized COVID-19 patients in China. *Sci. Rep.* **2020**, *10*, 15058. [CrossRef] [PubMed]
16. Vaughn, V.M.; Gandhi, T.N.; Petty, L.A.; Patel, P.K.; Prescott, H.C.; Malani, A.N.; Ratz, D.; McLaughlin, E.; Chopra, V.; Flanders, S.A. Empiric Antibacterial Therapy and Community-onset Bacterial Coinfection in Patients Hospitalized With Coronavirus Disease 2019 (COVID-19): A Multi-hospital Cohort Study. *Clin. Infect. Dis.* **2021**, *72*, e533–e541. [CrossRef]
17. Chang, M.; Dietz, D.; Shoucri, S.; Laracy, J.; Sobieszczyk, M.E.; Uhlemann, A.-C.; Zucker, J.; Kubin, C.J. Limited Utility of Procalcitonin in Identifying Community-Associated Bacterial Infections in Patients Presenting with Coronavirus Disease 2019. *Antimicrob. Agents Chemother.* **2021**, *65*, e02167-20. [CrossRef]
18. Mertens, P.; De Vos, N.; Martiny, D.; Vandenberg, O. Development and potential usefulness of the COVID-19 Ag Respi-Strip diagnostic assay in a pandemic context. *Front. Med.* **2020**, *7*, 225. [CrossRef]
19. Woodhead, M.; Blasi, F.; Ewig, S.; Garau, J.; Huchon, G.; Ieven, M.; Ortqvist, A.; Schaberg, T.; Torres, A.; van der Heijden, G.; et al. Guidelines for the management of adult lower respiratory tract infections-Full version. *Clin. Microbiol. Infect.* **2011**, *17*, E1–E59. [CrossRef]
20. Schuetz, P.; Wirz, Y.; Sager, R.; Christ-Crain, M.; Stolz, D.; Tamm, M.; Bouadma, L.; Luyt, C.E.; Wolff, M.; Chastre, J.; et al. Procalcitonin to initiate or discontinue antibiotics in acute respiratory tract infections. *Cochrane Database Syst. Rev.* **2017**, *2017*, CD007498. [CrossRef]
21. Hausfater, P.; Garric, S.; Ayed, S.B.; Rosenheim, M.; Bernard, M.; Riou, B. Usefulness of Procalcitonin as a Marker of Systemic Infection in Emergency Department Patients: A Prospective Study. *Clin. Infect. Dis.* **2002**, *34*, 895–901. [CrossRef]
22. Piacentini, E.; Sánchez, B.; Arauzo, V.; Calbo, E.; Cuchi, E.; Nava, J.M. Procalcitonin levels are lower in intensive care unit patients with H1N1 influenza A virus pneumonia than in those with community-acquired bacterial pneumonia. A pilot study. *J. Crit. Care* **2011**, *26*, 201–205. [CrossRef]
23. Cuquemelle, E.; Soulis, F.; Villers, D.; Roche-Campo, F.; Ara Somohano, C.; Fartoukh, M.; Kouatchet, A.; Mourvillier, B.; Dellamonica, J.; Picard, W.; et al. Can procalcitonin help identify associated bacterial infection in patients with severe influenza pneumonia? A multicentre study. *Intensive Care Med.* **2011**, *37*, 796–800. [CrossRef] [PubMed]
24. Siljan, W.W.; Holter, J.C.; Michelsen, A.E.; Nymo, S.H.; Lauritzen, T.; Oppen, K.; Husebye, E.; Ueland, T.; Mollnes, T.E.; Aukrust, P.; et al. Inflammatory biomarkers are associated with aetiology and predict outcomes in community-acquired pneumonia: Results of a 5-year follow-up cohort study. *ERJ Open Res.* **2019**, *5*, 00014-2019. [CrossRef]
25. Metlay, J.P.; Waterer, G.W.; Long, A.C.; Anzueto, A.; Brozek, J.; Crothers, K.; Cooley, L.A.; Dean, N.C.; Fine, M.J.; Flanders, S.A.; et al. Diagnosis and treatment of adults with community-acquired pneumonia. *Am. J. Respir. Crit. Care Med.* **2019**, *200*, E45–E67. [CrossRef] [PubMed]
26. Kalil, A.C.; Metersky, M.L.; Klompas, M.; Muscedere, J.; Sweeney, D.A.; Palmer, L.B.; Napolitano, L.M.; O'Grady, N.P.; Bartlett, J.G.; Carratalà, J.; et al. Executive Summary: Management of Adults With Hospital-acquired and Ventilator-associated Pneumonia: 2016 Clinical Practice Guidelines by the Infectious Diseases Society of America and the American Thoracic Society. *Clin. Infect. Dis.* **2016**, *63*, 575–582. [CrossRef]
27. Jain, S.; Self, W.H.; Wunderink, R.G.; Fakhran, S.; Balk, R.; Bramley, A.M.; Reed, C.; Grijalva, C.G.; Anderson, E.J.; Courtney, D.M.; et al. Community-acquired pneumonia requiring hospitalization among U.S. adults. *N. Engl. J. Med.* **2015**, *373*, 415–427. [CrossRef] [PubMed]
28. Chen, X.; Yang, Y.; Huang, M.; Liu, L.; Zhang, X.; Xu, J.; Geng, S.; Han, B.; Xiao, J.; Wan, Y. Differences between COVID-19 and suspected then confirmed SARS-CoV-2-negative pneumonia: A retrospective study from a single center. *J. Med. Virol.* **2020**, *92*, 1572–1579. [CrossRef]

29. Han, H.; Ma, Q.; Li, C.; Liu, R.; Zhao, L.; Wang, W.; Zhang, P.; Liu, X.; Gao, G.; Liu, F.; et al. Profiling serum cytokines in COVID-19 patients reveals IL-6 and IL-10 are disease severity predictors. *Emerg. Microbes Infect.* **2020**, *9*, 1123–1130. [CrossRef]
30. Ming, D.K.; Myall, A.C.; Hernandez, B.; Weiße, A.Y.; Peach, R.L.; Barahona, M.; Rawson, T.M.; Holmes, A.H. Informing antimicrobial management in the context of COVID-19: Understanding the longitudinal dynamics of C-reactive protein and procalcitonin. *BMC Infect. Dis.* **2021**, *21*, 932. [CrossRef]

antibiotics

MDPI

Review

Azithromycin through the Lens of the COVID-19 Treatment

Georgia G. Kournoutou * and George Dinos *

Department of Biochemistry, School of Medicine, University of Patras, 26504 Patras, Greece
* Correspondence: gkurnutu@upatras.gr (G.G.K.); dinosg@upatras.gr (G.D.); Tel.: +30-2610996259 (G.D.)

Abstract: Azithromycin has become famous in the last two years, not for its main antimicrobial effect, but for its potential use as a therapeutic agent for COVID-19 infection. Initially, there were some promising results that supported its use, but it has become clear that scientific results are insufficient to support such a positive assessment. In this review we will present all the literature data concerning the activity of azithromycin as an antimicrobial, an anti-inflammatory, or an antivirus agent. Our aim is to conclude whether its selection should remain as a valuable antivirus agent or if its use simply has an indirect therapeutic contribution due to its antimicrobial and/or immunomodulatory activity, and therefore, if its further use for COVID-19 treatment should be interrupted. This halt will prevent further antibiotic resistance expansion and will keep azithromycin as a valuable anti-infective therapeutic agent.

Keywords: macrolides; azithromycin; virus; coronavirus; COVID-19; immunolides; antivirus

Citation: Kournoutou, G.G.; Dinos, G. Azithromycin through the Lens of the COVID-19 Treatment. *Antibiotics* **2022**, *11*, 1063. https://doi.org/10.3390/antibiotics11081063

Academic Editors: Sabine Danielle Allard and Johan Van Laethem

Received: 29 June 2022
Accepted: 1 August 2022
Published: 5 August 2022

1. Introduction

Azithromycin (Azi) belongs to the large family of macrolide antibiotics, an important class of first-line antimicrobial agents [1]. Azi belongs to the second generation of macrolides, as a semisynthetic derivative of erythromycin with a modified macrolactone ring with 15 members instead of 14 members as in erythromycin (Figure 1). Although Azi did not exhibit improved activity against Gram-positive bacteria compared to the mother compound erythromycin [2–4], it was selected for further development due to its enhanced pharmacokinetic profiles. In particular, it was selected for its high half-life time and the ability to accumulate at high levels within lung tissue [5–9]. Clarithromycin (Figure 1) is another key second-generation 14-membered macrolide with similar features and structure, and while it was initially included in a few trial schemes as a potential therapeutic drug for COVID-19, it was rapidly discontinued [10,11]. Azi, like most of the other macrolides, is not only known for its antimicrobial activity, but it also has additional actions as either anti-inflammatory or antivirus agents. In this review we will present a summary of the existing literature data concerning azithromycin and will explain why it was initially hypothesized to have activity for COVID-19 treatment. Additionally, all studies on the use or non-use of azithromycin in the treatment of COVID-19 will be presented. Finally, we will discuss why Azi is not included anymore in therapeutic protocols and why its use must be interrupted to avoid increasing Azi pathogen resistance thereby maintaining the antibiotic as a useful therapeutic weapon for a longer time.

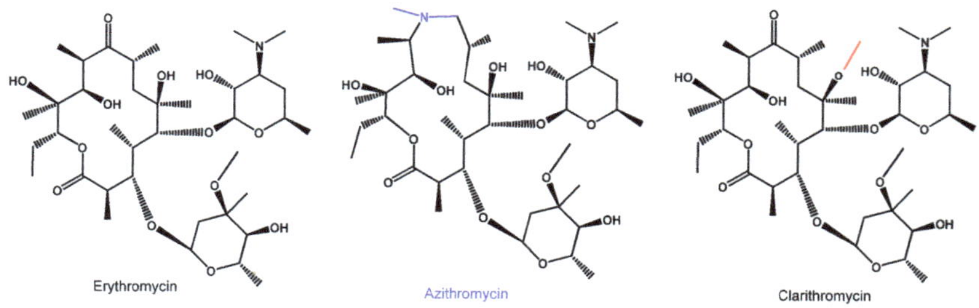

Figure 1. Molecular structure of the mother macrolide molecule erythromycin and its semisynthetic derivatives azithromycin (15-membered) and clarithromycin (14-membered). Blue and red colors in the structures represent modifications of the mother molecule (black).

2. Methods

The authors searched the PubMed and the Scopus database using the terms "Coronavirus", "COVID-19", and "Azithromycin". The same search terms were used for searching the Clinical Trials database. The abstracts were screened and only the relevant articles were considered in the review. Articles from the Clinical Trials database that were limited by a small sample size or other criteria were declared not relevant, and therefore, were not considered. Articles published as recently as 15 April 2022 were included.

3. Azithromycin as an Antimicrobial Agent

The antimicrobial activity of azithromycin results from its binding with high affinity to the entrance of the ribosomal exit tunnel of prokaryotic 70S ribosomes and strongly inhibiting the bacterial protein synthesis [12–14]. According to the crystal structure data, it binds to the entrance of the nascent peptide exit tunnel and partially occludes it (Figure 2).

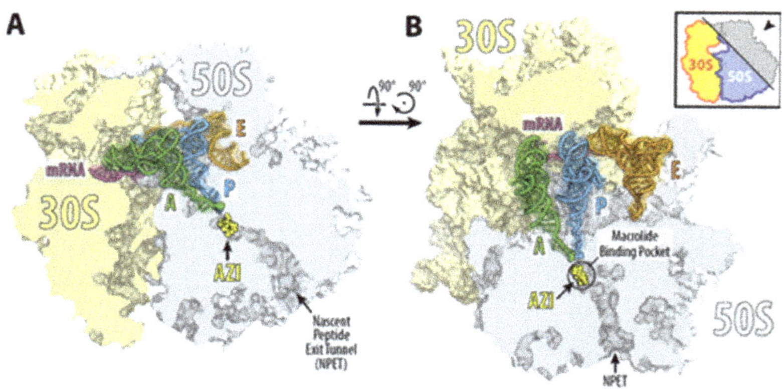

Figure 2. Structure of azithromycin in complex with the 70S ribosome carrying A-, P-, and E-site tRNAs. (**A**,**B**) Location of the ribosome-bound azithromycin (yellow) in the macrolide binding pocket at the entrance to the nascent peptide exit tunnel (NPET) of the 70S ribosome relative to tRNAs viewed as cross-cut sections through the ribosome. The 30S subunit is shown in light yellow, the 50S subunit is in light blue, the mRNA is in magenta, and the A-, P-, and E-site tRNAs are colored green, dark blue, and orange, respectively. The phenylalanyl and formyl-methionyl moieties of the A- and P-site tRNAs are shown as spheres [15].

Thus, Azi was considered as 'tunnel plugs' that inhibit the synthesis of every protein entering the exit tunnel [1,16]. However, more recent evidence demonstrates that macrolides selectively inhibit the translation of a subset of cellular proteins and that their action crucially depends on the nascent protein sequence and the antibiotic structure [17].

Recent studies have shown that the translation of many genes was arrested at a few distinct sites through the length of the gene after treatment with macrolide antibiotics. Analysis of the sites of the stops revealed the existence of specific sequence signatures that induce pronounced drug-induced translation arrest and lead to specific regulation of protein synthesis. Ribo-seq and toeprinting experiments have revealed leader ORFs of macrolide resistance genes carrying the +x+ motif, where + stands for positively charged amino acids lysine or arginine, and x stands for any amino acid [17,18].

Therefore, Azi emerges as a modulator of translation rather than as a global inhibitor of protein synthesis. In general, macrolide antibiotics are active mainly against Gram-positive bacteria and have a lower activity against Gram-negative bacteria [19,20]. Macrolides are very active against Gram-positive bacteria *Staphylococcus*, *Streptococcus*, and *Diplococcus*; among Gram-negative cocci, *Neisseria gonorrhea*, *Haemophilus influenzae*, *Bordetella pertussis*, and *Neisseria meningitis*; and are extremely active against various Mycoplasmas. Since its discovery Azi has been extensively used in the treatment of bacterial and mycobacterial infections of the respiratory, gastrointestinal, genitourinary, and cutaneous systems [21–23]. Azi is a member of the WHO list of essential medications [24] and is available in large quantities worldwide. Despite some mild side effects, including mainly diarrhea and QT prolongation, Azi is proven to be safe and cheap, and therefore, easily available to humans worldwide [21,22].

4. Azithromycin as an Anti-Inflammatory Agent

Beyond the antibacterial activity of azithromycin, and broadly most macrolides, their anti-inflammatory effects have been established and some of them have been used in chronic inflammatory diseases such as chronic rhinosinusitis, bronchial asthma, bronchiectasis, chronic obstructive pulmonary disease, cystic fibrosis, etc. [23–27]. Probably, the most striking example of their immunomodulation comes from diffuse panbronchiolitis, an idiopathic inflammation and progressively destructive disease of the bronchioles which can be converted from a lethal to a treatable disease with daily low-dose erythromycin or Azi [23,28,29]. This has been accredited to the ability of Azi to normalize the upregulated activities of IL-1β, IL-2, TNF, and GM-CSF [30]. Azi is rapidly absorbed after oral administration with a half-life time of approximately 3 days, leading to a high and constant tissue concentration [23,31]. As a result, Azi accumulates in human cells, including epithelial cells, and most notably in phagocytes where it has been concentrated hundreds to thousands of times with a focus on phagocyte lysosomes [9,31]. Its anti-inflammatory or immunomodulatory activity reported in several studies includes the most frequent effects on neutrophils, monocytes, and lymphocytes [27,29,32]. Among the usually measured immunological modified markers are the number of decreased neutrophils; the concentrations of neutrophil elastase; cytokines release; surface-expressed molecules (mainly Toll-like receptors); superoxide production; and cell homeostasis, mainly apoptosis and phagocytosis (Figure 3) [27,29,32,33]. Neutrophil function inhibition has been reported more frequently than eosinophil function inhibition. Azi stimulates neutrophil degranulation and phagocytosis-associated oxidative burst, mediated via modulation of ERK 1/2 signaling [19]. These initial stimulatory effects are followed by modulation of transcription factors activator protein (AP)-1, nuclear factor kappa B (NFκB), inflammatory cytokines, and mucin release, with overall anti-inflammatory effects [34].

Azi inhibits lipopolysaccharide-induced pro-inflammatory cytokines; increases phagocytosis by inhibiting AP-1 [35]; improves lysosomal resistance to oxidant challenge [36]; and promotes M2 polarization of macrophages (a process in which macrophages produce distinct functional phenotypes in response to specific microenvironmental stimuli and signals) [37–39]. Azi can also increase the phagocytosis of apoptotic epithelial cells [40] and neutrophils by macrophages [41] further supporting its anti-inflammatory activity. Studies have shown that part of the immunomodulatory effects of macrolides could be attributed to the impairment of TLR signaling by reducing the release of PAMPs (Pathogen-Associated

Molecular Patterns) and inhibiting TLR expression, either of dendritic cells or macrophages thereby regulating the immune response [33,42].

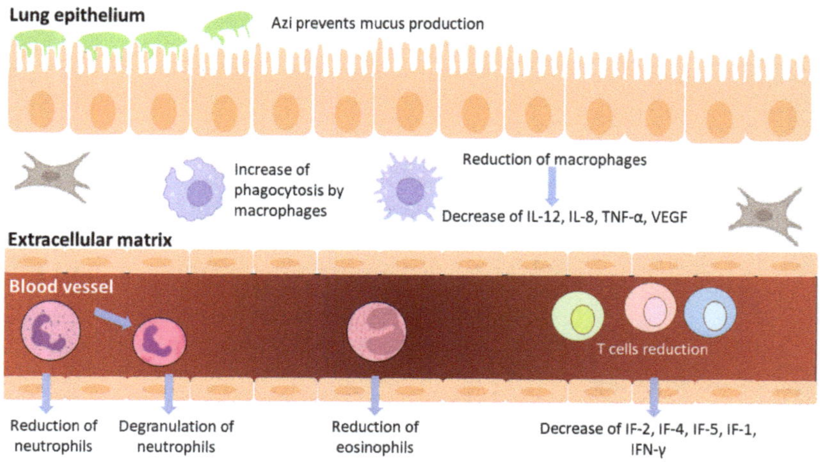

Figure 3. Immunomodulatory effects of azithromycin.

Immunomodulatory effects, although similar to most therapeutic macrolides, are likely to differ among them. Few studies have examined the anti-inflammatory effects of macrolides on more than one macrolide, and none of the human trials have explicitly compared different macrolides. Furthermore, the majority of these trials were conducted on healthy volunteers and/or Azi was administered in varying doses at a time [27,43–45]. Clinical investigations in CF patients, on the other hand, revealed that Azi, but not Clarithromycin, improves respiratory function and reduces pulmonary exacerbations [46,47]. Additionally, another study showed that Azi, but not clarithromycin or roxithromycin, inhibits IL-1alpha and IL-1beta production [48]. In general, azithromycin inhibits the synthesis of pro-inflammatory cytokines by both innate and adaptive immune cells, as well as the accumulation, adhesion, and death of pulmonary neutrophils [32].

Azithromycin, like other macrolides, has very low activity against eukaryotes due to their low affinity for binding to eukaryotic ribosomes [1]. There are specific differences between eukaryotic and bacterial ribosomes (differences between rRNA bases or ribosomal proteins) that mediate the selectivity and toxicity of ribosomal drugs, as established by rRNA sequencing studies and X-ray crystallography [49].

5. Azithromycin as an Antivirus Agent

Azi's antiviral effects have been demonstrated in vitro, albeit not all examples have been confirmed in vivo [32]. Since Azi mediated exacerbations in airway diseases, particularly in asthma [25,50], its effects were studied against viruses that cause such airway infections such as rhinoviruses (RV). Azi inhibits RV replication and releases in primary human bronchial epithelial cells in vitro [51]. The AMAZES research, the largest clinical trial of a long-term macrolide on airway diseases, found that Azi reduced asthma exacerbations by 40% in vivo [25]. The mechanism is not known, but metagenomic analysis suggested that it could be related to an antibacterial effect versus *Haemophilus influenzae* and possibly its abundance in inhaled air [52–54]. Pre-treatment with azithromycin inhibits RV replication in CF bronchial epithelial cells, probably by amplifying the antiviral response mediated by the IFN pathway [55]. Additionally, Azi showed a reduction in H1N1 viral replication in A549 cells with IC_{50} 58 μM interfering with the internalization of viruses [56]. In experiments with the Zika virus, within glial cell lines and human astrocytes, there was a reduction in viral growth and virus-induced cytotoxicity [57]. Equally, Azi inhibited Ebola replication with EC_{50} 5.1 μM and low toxicity; however, it did not boost survival in mice or guinea pigs when tested in vivo in a mouse model [58].

The precise mechanism of the antiviral activity of Azi remains unclear. Given that Azi is a weak base it can accumulate in acidic intracellular organelles such as endosomal vesicles and lysosomes [59]. In keeping with lysosomal accumulation, azithromycin causes lysosomal pH change [23]. This modified acidic environment caused by accumulation of Azi could also be responsible for uncoating enveloped viruses such as influenza and maybe coronavirus [59]. Data also suggest that the antiviral activity of Azi could be attributed to its ability to increase the expression of the epithelial interferon genes, leading to a reduction in viral replication [60].

Recently, Azi and spiramycin (a natural 16-membered ring macrolide) provided significant in vivo protection against enterovirus-A71 infection in mice [61]. Spiramycin was found to interfere with EV-A71 viral RNA synthesis, and it is likely that spiramycin and Azi function in concert after the viral entrance; thereby, inhibiting viral RNA synthesis either directly or indirectly.

6. Azithromycin and Betacoronovirus

From the beginning of the current SARS-CoV-2 pandemic, several drug screens were conducted in a rapid, urgent manner to evaluate potential candidate medications against this pathogen. The requirements were: to be approved, to be inexpensive, to be safe, and to be available as quickly as was feasible worldwide. Previous screens had recognized more than 90 drugs that inhibited SARS-CoV-2 viral replication with EC_{50} nearly to 10 μM [62]. The tested drugs included protease inhibitors, ATPase proton pump inhibitors, viral protease inhibitors, compounds targeting the angiotensin pathway, and antibiotics. Azi tested in Vero E6 cells had an EC_{50} of 2.12 μM and EC_{90} equal to 8.65 μM, and selectivity index >19 [63], which is very comparable to the control antiviral-compound remdesivir (EC_{50} = 1.65, EC_{90} = 2.52), the first antiviral agent with proven clinical efficacy against SARS-CoV-2 in all clinical trials [64–66]. Azi was also discovered as a target in a bioinformatic screening investigation of potentially relevant pathways that may be turned into pharmaceutically acceptable forms. An initial study focused on two of the previous candidate molecules, hydroxychloroquine (HQL) and Azi, suggested a synergistic inhibition of SARS-CoV-2 replication in Vero cells at 5 and 10 μM concentrations, respectively [67,68]. This synergy was presented as a way to make hydroxychloroquine more effective at less hazardous concentrations. It was the first observational study suggesting that HQL, especially when combined with Azi, improved virological clearance [69]. However, because the data with Azi came from only six patients and the study was open-label and nonrandomized, no acceptable conclusions could be derived statistically [70]. This Azi-HQL combination was also investigated in nonhuman primates, but no substantial antiviral effect was observed in the five macaques given Azi in addition to hydroxychloroquine [71]. Furthermore, this initial favorable finding led to the immediate start of interventional trials to assess the efficacy of the COVID-19 therapy combination, as well as the efficacy of Azi with HQL. Hundreds of trials with Azi are listed on clinical trials.gov. Initially, Azi was prescribed as an adjunct to hydroxychloroquine, but later HQL was largely abandoned and Azi was used alone. From the beginning of 2020, decades of publications were released either favoring or discouraging the use of Azi, both with or without HQL [70]. According to them, Azi was initially favored with or without HQL [72–77] but at the same time more observations did not favor its use [78–86]. Since most of them were retrospective studies, it was clear that randomized control trials (RCTs) were necessary to clarify the previous controversial data. All these RCTs were integrated during the previous year and are presented in Table 1. The table gives an overview of the most currently published, up-to-date, peer-reviewed studies in the literature, in which the effect of Azi is evaluated. Although these RCTs in Table 1 differ in their outcomes and whether or not hospitalized patients are included, all of them suggested that azithromycin does not reduce hospital admissions, respiratory failure, or death when compared to conventional therapy, and therefore, Azi should no longer be used to treat COVID-19. In a few words, all of them showed that, in hospitalized patients with COVID-19, azithromycin did not reduce the time to sustained clinical improvement or discharge. There is clearly no efficacy in terms of clinical status or mortality at the fixed time points used in all scientifically acceptable large trials.

Table 1. Published RCTs assessing the effect of Azithromycin on COVID-19 treatment.

Name of Clinical Trial	Year	Participants	Clinical Outcome	Type of Study
Furtado et al., 2020 (COALITION II) [87]	2020	447	No justification for the use of azithromycin for better clinical outcomes.	Open label
Butler et al., 2021 (PRINCIPLE) [88]	2021	292	No justification for the use of azithromycin for reducing time to recovery or risk of hospitalization.	Open label
Hinks et al., 2021 (ATOMIC 2) [89]	2021	263	No justification for the use of azithromycin for reducing the risk of hospitalization or death.	Open label
Horby et al., 2021 (RECOVERY) [90]	2021	2265	No justification for the use of azithromycin on inpatients for increase in survival.	Open label
Oldenburg et al., 2021 [11]	2021	7763	No justification for the use of azithromycin versus placebo for elimination of symptoms at day 14.	Blind
Gyselinck et al., 2022 (DAWn-AZITHRO) [70]	2022	160	Early trial termination, failed to demonstrate a benefit of azithromycin.	Open label

Furthermore, according to Oldenburg et al., there was no significant difference in self-reported symptom absence 14 days after enrollment among patients assigned to azithromycin versus a placebo in their randomized controlled trial of single-dose oral azithromycin for outpatient COVID-19 [11]. This last finding supports earlier randomized clinical trials of azithromycin for COVID-19 in both outpatient and inpatient settings, none of which found azithromycin to be effective in treating COVID-19.

Given that azithromycin consumption during the pandemic was increased up to 3 times compared to the pre COVID period [91–93], it is important to reduce useless consumption, as it is an extremely dangerous practice, to avoid increasing antimicrobial resistance (AMR). Antimicrobial resistance (AMR) develops when bacteria, fungi, or viruses are exposed to antibiotics, antifungals, or antivirals leading to the development of a resistance to one or more antimicrobial drugs. As a result, the antimicrobials become ineffective and infections may persist. AMR is considered a serious and persistent therapeutic problem today being an economic and health burden. It is conservatively estimated that, in the US and Europe, 2.5 million people are affected by such infections each year and approximately 50,000 people die because of these infections [94]. The discovery of novel antibiotics has nearly halted over the past 30 years leading to the exhaustion of the pipeline reserve. The resistance of pathogens to antibiotics can be addressed with a rapid development of new effective and safe antibiotics [1,95]. Several studies have revealed a significant increase in drug resistance to azithromycin in some strains of *gonococci* [96]. Drug resistance to azithromycin is also increasing in *E. coli* [95].

Identifying strategies that can work to reduce the burden of bacterial AMR—either across a wide range of settings or those that are specifically tailored to the resources available and leading pathogen–drug combinations in a particular setting—is an urgent priority [97]. Since the prevalence of bacterial superinfection in COVID-19 is low [98], and unlike influenza [99–101], there is no preventive benefit against postviral pneumococcal and atypical pneumonia [98], it will be extremely helpful to avoid the useless consumption of any antibiotic prescription, specifically azithromycin, in COVID-19 treatment..

The outbreak of a pandemic led to a massive disruption of healthcare systems which overshadowed the misuse and incorrect prescription of some antibiotics [102]. Many COVID-19 patients received empirical antibiotic therapy for COVID 19 treatment in the early stages of pandemic since it was considered the safer option due to clinical uncertainty [93]. Control policies should be administrated in clinical practice regarding the use of drugs in the treatment of COVID-19. As AMR will be a major clinical problem, stewardship activities are necessary in the coming years to face the new-pandemic [102].

7. Closing Remarks and Perspectives

During the pandemic of coronavirus, antibiotics prescription was elevated without justification, partly because the medical community was unprepared for this burst and secondly because the clinical situation of patients changed dramatically each day after the initial day of infection. The administration of known antibiotics was considered to be the correct way of combating coronavirus but it soon became clear that there was no justification for the overuse of antibiotics as they did not decrease the risk of mortality in the patients who had no reason to receive this treatment [103].

To summarize, there is no scientific justification for the use of azithromycin in the treatment of COVID-19 up to now, and the only way to keep this antibiotic relevant in the future as a useful tool for combating pathogenic infections is to use it wisely, only after careful consideration and high expectations.

Author Contributions: G.G.K.: writing—review and editing and writing—original draft; G.D.: writing—review and editing. All authors have read and agreed to the published version of the manuscript.

Funding: The publication of this article has been financed by the Research Committee of the University of Patras. G.G.K. is a Research Fellow supported by the Hellenic State Scholarships Foundation (IKY) through the Operational Program "Human Resources Development, Education and Lifelong Learning" co-financed by Greece and the European Union (European Social Fund, ESF).

Institutional Review Board Statement: Not applicable.

Informed Consent Statement: Not applicable.

Data Availability Statement: All relevant data are presented in the article.

Acknowledgments: The authors want to thank Yuri Polikanov for the kind preparation of Figure 2.

Conflicts of Interest: The authors declare no conflict of interest.

Abbreviations

Azi: azithromycin; IL-1β: interleukin 1β; IL-2: interleukin 2; TNF: tissue necrosis factor; GM-CSF: granulocyte macrophage colony stimulating factor; TLR: Toll-like receptor; AP-1: activator protein-1; EV-A71: enterovirus A71; RV: rhinovirus; CF; cystic fibrosis; IFN: interferon; HQL: hydroxychloroquine; ERK1/2: extracellular signal related regulated kinase 1/2.

References

1. Dinos, G.P. The Macrolide Antibiotic Renaissance. *Br. J. Pharmacol.* **2017**, *174*, 2967–2983. [CrossRef] [PubMed]
2. Barry, A.L.; Jones, R.N.; Thornsberry, C. In Vitro Activities of Azithromycin (CP 62,993), Clarithromycin (A-56268; TE-031), Erythromycin, Roxithromycin, and Clindamycin. *Antimicrob. Agents Chemother.* **1988**, *32*, 752–754. [CrossRef] [PubMed]
3. Barry, A.L.; Fuchs, P.C.; Brown, S.D. Relative Potency of Telithromycin, Azithromycin and Erythromycin against Recent Clinical Isolates of Gram-Positive Cocci. *Eur. J. Clin. Microbiol. Infect. Dis. Off. Publ. Eur. Soc. Clin. Microbiol.* **2001**, *20*, 494–497. [CrossRef] [PubMed]
4. Fernandes, P.B.; Hardy, D.J. Comparative in Vitro Potencies of Nine New Macrolides. *Drugs Exp. Clin. Res.* **1988**, *14*, 445–451. [PubMed]
5. Wise, R. The Development of Macrolides and Related Compounds. *J. Antimicrob. Chemother.* **1989**, *23*, 299–300. [CrossRef]

6. Foulds, G.; Shepard, R.M.; Johnson, R.B. The Pharmacokinetics of Azithromycin in Human Serum and Tissues. *J. Antimicrob. Chemother.* **1990**, *25* (Suppl. A), 73–82. [CrossRef]
7. Retsema, J.A.; Girard, A.E.; Girard, D.; Milisen, W.B. Relationship of High Tissue Concentrations of Azithromycin to Bactericidal Activity and Efficacy In Vivo. *J. Antimicrob. Chemother.* **1990**, *25* (Suppl. A), 83–89. [CrossRef]
8. Hardy, D.J.; Guay, D.R.P.; Jones, R.N. Clarithromycin, a Unique Macrolide. A Pharmacokinetic, Microbiological, and Clinical Overview. *Diagn. Microbiol. Infect. Dis.* **1992**, *15*, 39–53. [CrossRef]
9. Wildfeuer, A.; Laufen, H.; Zimmermann, T. Distribution of Orally Administered Azithromycin in Various Blood Compartments. *Int. J. Clin. Pharmacol. Ther.* **1994**, *32*, 356–360.
10. Ayerbe, L.; Risco-Risco, C.; Forgnone, I.; Pérez-Piñar, M.; Ayis, S. Azithromycin in Patients with COVID-19: A Systematic Review and Meta-Analysis. *J. Antimicrob. Chemother.* **2022**, *77*, 303–309. [CrossRef]
11. Oldenburg, C.E.; Pinsky, B.A.; Brogdon, J.; Chen, C.; Ruder, K.; Zhong, L.; Nyatigo, F.; Cook, C.A.; Hinterwirth, A.; Lebas, E.; et al. Effect of Oral Azithromycin vs Placebo on COVID-19 Symptoms in Outpatients with SARS-CoV-2 Infection: A Randomized Clinical Trial. *JAMA* **2021**, *326*, 490–498. [CrossRef] [PubMed]
12. Schlünzen, F.; Zarivach, R.; Harms, J.; Bashan, A.; Tocilj, A.; Albrecht, R.; Yonath, A.; Franceschi, F. Structural Basis for the Interaction of Antibiotics with the Peptidyl Transferase Centre in Eubacteria. *Nature* **2001**, *413*, 814–821. [CrossRef] [PubMed]
13. Bulkley, D.; Innis, C.A.; Blaha, G.; Steitz, T.A. Revisiting the Structures of Several Antibiotics Bound to the Bacterial Ribosome. *Proc. Natl. Acad. Sci. USA* **2010**, *107*, 17158–17163. [CrossRef]
14. Dunkle, J.A.; Xiong, L.; Mankin, A.S.; Cate, J.H.D. Structures of the Escherichia Coli Ribosome with Antibiotics Bound near the Peptidyl Transferase Center Explain Spectra of Drug Action. *Proc. Natl. Acad. Sci. USA* **2010**, *107*, 17152–17157. [CrossRef]
15. Svetlov, M.S.; Plessa, E.; Chen, C.W.; Bougas, A.; Krokidis, M.G.; Dinos, G.P.; Polikanov, Y.S. High-Resolution Crystal Structures of Ribosome-Bound Chloramphenicol and Erythromycin Provide the Ultimate Basis for Their Competition. *RNA* **2019**, *25*, 600–606. [CrossRef] [PubMed]
16. Kannan, K.; Kanabar, P.; Schryer, D.; Florin, T.; Oh, E.; Bahroos, N.; Tenson, T.; Weissman, J.S.; Mankin, A.S. The General Mode of Translation Inhibition by Macrolide Antibiotics. *Proc. Natl. Acad. Sci. USA* **2014**, *111*, 15958–15963. [CrossRef]
17. Vázquez-Laslop, N.; Mankin, A.S. How Macrolide Antibiotics Work. *Trends Biochem. Sci.* **2018**, *43*, 668–684. [CrossRef]
18. Beckert, B.; Leroy, E.C.; Sothiselvam, S.; Bock, L.V.; Svetlov, M.S.; Graf, M.; Arenz, S.; Abdelshahid, M.; Seip, B.; Grubmüller, H.; et al. Structural and Mechanistic Basis for Translation Inhibition by Macrolide and Ketolide Antibiotics. *Nat. Commun.* **2021**, *12*, 4466. [CrossRef]
19. Ishimoto, H.; Mukae, H.; Sakamoto, N.; Amenomori, M.; Kitazaki, T.; Imamura, Y.; Fujita, H.; Ishii, H.; Nakayama, S.; Yanagihara, K.; et al. Different Effects of Telithromycin on MUC5AC Production Induced by Human Neutrophil Peptide-1 or Lipopolysaccharide in NCI-H292 Cells Compared with Azithromycin and Clarithromycin. *J. Antimicrob. Chemother.* **2009**, *63*, 109–114. [CrossRef]
20. Nakayama, I. Macrolides in clinical practice. In *Macrolide Antibiotics: Chemistry, Biology and Practice*, 1st ed.; Omura, S., Ed.; Academic Press: Orlando, FL, USA, 2022.
21. Firth, A.; Prathapan, P. Azithromycin: The First Broad-Spectrum Therapeutic. *Eur. J. Med. Chem.* **2020**, *207*, 112739. [CrossRef]
22. Taylor, W.R.; Richie, T.L.; Fryauff, D.J.; Ohrt, C.; Picarima, H.; Tang, D.; Murphy, G.S.; Widjaja, H.; Braitman, D.; Tjitra, E.; et al. Tolerability of Azithromycin as Malaria Prophylaxis in Adults in Northeast Papua, Indonesia. *Antimicrob. Agents Chemother.* **2003**, *47*, 2199–2203. [CrossRef] [PubMed]
23. Parnham, M.J.; Erakovic Haber, V.; Giamarellos-Bourboulis, E.J.; Perletti, G.; Verleden, G.M.; Vos, R. Azithromycin: Mechanisms of Action and Their Relevance for Clinical Applications. *Pharmacol. Ther.* **2014**, *143*, 225–245. [CrossRef] [PubMed]
24. WHO. WHO Model List of Essential Medicines—22nd List. 2021. Available online: https://www.who.int/publications/i/item/WHO-MHP-HPS-EML-2021.02. (accessed on 15 February 2022).
25. Gibson, P.G.; Yang, I.A.; Upham, J.W.; Reynolds, P.N.; Hodge, S.; James, A.L.; Jenkins, C.; Peters, M.J.; Marks, G.B.; Baraket, M.; et al. Effect of Azithromycin on Asthma Exacerbations and Quality of Life in Adults with Persistent Uncontrolled Asthma (AMAZES): A Randomised, Double-Blind, Placebo-Controlled Trial. *Lancet (Lond. Engl.)* **2017**, *390*, 659–668. [CrossRef]
26. Hansen, M.P.; Scott, A.M.; McCullough, A.; Thorning, S.; Aronson, J.K.; Beller, E.M.; Glasziou, P.P.; Hoffmann, T.C.; Clark, J.; Del Mar, C.B. Adverse Events in People Taking Macrolide Antibiotics versus Placebo for Any Indication. *Cochrane Database Syst. Rev.* **2019**, *18*, CD011825. [CrossRef] [PubMed]
27. Zimmermann, P.; Ziesenitz, V.C.; Curtis, N.; Ritz, N. The Immunomodulatory Effects of Macrolides—A Systematic Review of the Underlying Mechanisms. *Front. Immunol.* **2018**, *9*, 302. [CrossRef] [PubMed]
28. Kudoh, S.; Azuma, A.; Yamamoto, M.; Izumi, T.; Ando, M. Improvement of Survival in Patients with Diffuse Panbronchiolitis Treated with Low-Dose Erythromycin. *Am. J. Respir. Crit. Care Med.* **1998**, *157*, 1829–1832. [CrossRef]
29. Zarogoulidis, P.; Papanas, N.; Kioumis, I.; Chatzaki, E.; Maltezos, E.; Zarogoulidis, K. Macrolides: From in Vitro Anti-Inflammatory and Immunomodulatory Properties to Clinical Practice in Respiratory Diseases. *Eur. J. Clin. Pharmacol.* **2012**, *68*, 479–503. [CrossRef] [PubMed]
30. Weng, D.; Wu, Q.; Chen, X.-Q.; Du, Y.-K.; Chen, T.; Li, H.; Tang, D.-L.; Li, Q.-H.; Zhang, Y.; Lu, L.-Q.; et al. Azithromycin Treats Diffuse Panbronchiolitis by Targeting T Cells via Inhibition of MTOR Pathway. *Biomed. Pharmacother.* **2019**, *110*, 440–448. [CrossRef]

31. Wilms, E.B.; Touw, D.J.; Heijerman, H.G.M. Pharmacokinetics of Azithromycin in Plasma, Blood, Polymorphonuclear Neutrophils and Sputum during Long-Term Therapy in Patients with Cystic Fibrosis. *Ther. Drug Monit.* **2006**, *28*, 219–225. [CrossRef]
32. Oliver, M.E.; Hinks, T.S.C. Azithromycin in Viral Infections. *Rev. Med. Virol.* **2021**, *31*, e2163. [CrossRef]
33. Reijnders, T.D.Y.; Saris, A.; Schultz, M.J.; van der Poll, T. Immunomodulation by Macrolides: Therapeutic Potential for Critical Care. *Lancet. Respir. Med.* **2020**, *8*, 619–630. [CrossRef]
34. Araki, N.; Yanagihara, K.; Morinaga, Y.; Yamada, K.; Nakamura, S.; Yamada, Y.; Kohno, S.; Kamihira, S. Azithromycin Inhibits Nontypeable Haemophilus Influenzae-Induced MUC5AC Expression and Secretion via Inhibition of Activator Protein-1 in Human Airway Epithelial Cells. *Eur. J. Pharmacol.* **2010**, *644*, 209–214. [CrossRef] [PubMed]
35. Hodge, S.; Hodge, G.; Jersmann, H.; Matthews, G.; Ahern, J.; Holmes, M.; Reynolds, P.N. Azithromycin Improves Macrophage Phagocytic Function and Expression of Mannose Receptor in Chronic Obstructive Pulmonary Disease. *Am. J. Respir. Crit. Care Med.* **2008**, *178*, 139–148. [CrossRef] [PubMed]
36. Persson, H.L.; Vainikka, L.K.; Sege, M.; Wennerström, U.; Dam-Larsen, S.; Persson, J. Leaky Lysosomes in Lung Transplant Macrophages: Azithromycin Prevents Oxidative Damage. *Respir. Res.* **2012**, *13*, 83. [CrossRef]
37. Murphy, B.S.; Sundareshan, V.; Cory, T.J.; Hayes, D.J.; Anstead, M.I.; Feola, D.J. Azithromycin Alters Macrophage Phenotype. *J. Antimicrob. Chemother.* **2008**, *61*, 554–560. [CrossRef]
38. Legssyer, R.; Huaux, F.; Lebacq, J.; Delos, M.; Marbaix, E.; Lebecque, P.; Lison, D.; Scholte, B.J.; Wallemacq, P.; Leal, T. Azithromycin Reduces Spontaneous and Induced Inflammation in DeltaF508 Cystic Fibrosis Mice. *Respir. Res.* **2006**, *7*, 134. [CrossRef]
39. Yamauchi, K.; Shibata, Y.; Kimura, T.; Abe, S.; Inoue, S.; Osaka, D.; Sato, M.; Igarashi, A.; Kubota, I. Azithromycin Suppresses Interleukin-12p40 Expression in Lipopolysaccharide and Interferon-Gamma Stimulated Macrophages. *Int. J. Biol. Sci.* **2009**, *5*, 667–678. [CrossRef]
40. Hodge, S.; Hodge, G.; Brozyna, S.; Jersmann, H.; Holmes, M.; Reynolds, P.N. Azithromycin Increases Phagocytosis of Apoptotic Bronchial Epithelial Cells by Alveolar Macrophages. *Eur. Respir. J.* **2006**, *28*, 486–495. [CrossRef]
41. Yamaryo, T.; Oishi, K.; Yoshimine, H.; Tsuchihashi, Y.; Matsushima, K.; Nagatake, T. Fourteen-Member Macrolides Promote the Phosphatidylserine Receptor-Dependent Phagocytosis of Apoptotic Neutrophils by Alveolar Macrophages. *Antimicrob. Agents Chemother.* **2003**, *47*, 48–53. [CrossRef]
42. Huang, S.-W.; Chen, Y.-J.; Wang, S.-T.; Ho, L.-W.; Kao, J.-K.; Narita, M.; Takahashi, M.; Wu, C.-Y.; Cheng, H.-Y.; Shieh, J.-J. Azithromycin Impairs TLR7 Signaling in Dendritic Cells and Improves the Severity of Imiquimod-Induced Psoriasis-like Skin Inflammation in Mice. *J. Dermatol. Sci.* **2016**, *84*, 59–70. [CrossRef]
43. Culić, O.; Eraković, V.; Cepelak, I.; Barisić, K.; Brajsa, K.; Ferencić, Z.; Galović, R.; Glojnarić, I.; Manojlović, Z.; Munić, V.; et al. Azithromycin Modulates Neutrophil Function and Circulating Inflammatory Mediators in Healthy Human Subjects. *Eur. J. Pharmacol.* **2002**, *450*, 277–289. [CrossRef]
44. Criqui, G.I.; Solomon, C.; Welch, B.S.; Ferrando, R.E.; Boushey, H.A.; Balmes, J.R. Effects of Azithromycin on Ozone-Induced Airway Neutrophilia and Cytokine Release. *Eur. Respir. J.* **2000**, *15*, 856–862. [CrossRef] [PubMed]
45. Aubert, J.D.; Juillerat-Jeanneret, L.; Fioroni, P.; Dayer, P.; Plan, P.A.; Leuenberger, P. Function of Human Alveolar Macrophages after a 3-Day Course of Azithromycin in Healthy Volunteers. *Pulm. Pharmacol. Ther.* **1998**, *11*, 263–269. [CrossRef] [PubMed]
46. Southern, K.W.; Barker, P.M.; Solis-Moya, A.; Patel, L. Macrolide Antibiotics for Cystic Fibrosis. *Cochrane database Syst. Rev.* **2012**, *11*, CD002203. [CrossRef] [PubMed]
47. Robinson, P.; Schechter, M.S.; Sly, P.D.; Winfield, K.; Smith, J.; Brennan, S.; Shinkai, M.; Henke, M.O.; Rubin, B.K. Clarithromycin Therapy for Patients with Cystic Fibrosis: A Randomized Controlled Trial. *Pediatr. Pulmonol.* **2012**, *47*, 551–557. [CrossRef]
48. Gualdoni, G.A.; Lingscheid, T.; Schmetterer, K.G.; Hennig, A.; Steinberger, P.; Zlabinger, G.J. Azithromycin Inhibits IL-1 Secretion and Non-Canonical Inflammasome Activation. *Sci. Rep.* **2015**, *5*, 12016. [CrossRef]
49. Böttger, E.C.; Springer, B.; Prammananan, T.; Kidan, Y.; Sander, P. Structural Basis for Selectivity and Toxicity of Ribosomal Antibiotics. *EMBO Rep.* **2001**, *2*, 318–323. [CrossRef]
50. Brusselle, G.G.; Vanderstichele, C.; Jordens, P.; Deman, R.; Slabbynck, H.; Ringoet, V.; Verleden, G.; Demedts, I.K.; Verhamme, K.; Delporte, A.; et al. Azithromycin for Prevention of Exacerbations in Severe Asthma (AZISAST): A Multicentre Randomised Double-Blind Placebo-Controlled Trial. *Thorax* **2013**, *68*, 322–329. [CrossRef]
51. Gielen, V.; Johnston, S.L.; Edwards, M.R. Azithromycin Induces Anti-Viral Responses in Bronchial Epithelial Cells. *Eur. Respir. J.* **2010**, *36*, 646–654. [CrossRef]
52. Sajjan, U.S.; Jia, Y.; Newcomb, D.C.; Bentley, J.K.; Lukacs, N.W.; LiPuma, J.J.; Hershenson, M.B.H. Influenzae Potentiates Airway Epithelial Cell Responses to Rhinovirus by Increasing ICAM-1 and TLR3 Expression. *FASEB J. Off. Publ. Fed. Am. Soc. Exp. Biol.* **2006**, *20*, 2121–2123. [CrossRef]
53. Taylor, S.L.; Leong, L.E.X.; Mobegi, F.M.; Choo, J.M.; Wesselingh, S.; Yang, I.A.; Upham, J.W.; Reynolds, P.N.; Hodge, S.; James, A.L.; et al. Long-Term Azithromycin Reduces Haemophilus Influenzae and Increases Antibiotic Resistance in Severe Asthma. *Am. J. Respir. Crit. Care Med.* **2019**, *200*, 309–317. [CrossRef] [PubMed]
54. Taylor, S.L.; Ivey, K.L.; Gibson, P.G.; Simpson, J.L.; Rogers, G.B. Airway Abundance of Haemophilus Influenzae Predicts Response to Azithromycin in Adults with Persistent Uncontrolled Asthma. *Eur. Respir. J.* **2020**, *56*, 2000194. [CrossRef] [PubMed]
55. Schögler, A.; Kopf, B.S.; Edwards, M.R.; Johnston, S.L.; Casaulta, C.; Kieninger, E.; Jung, A.; Moeller, A.; Geiser, T.; Regamey, N.; et al. Novel Antiviral Properties of Azithromycin in Cystic Fibrosis Airway Epithelial Cells. *Eur. Respir. J.* **2015**, *45*, 428–439. [CrossRef] [PubMed]

56. Tran, D.H.; Sugamata, R.; Hirose, T.; Suzuki, S.; Noguchi, Y.; Sugawara, A.; Ito, F.; Yamamoto, T.; Kawachi, S.; Akagawa, K.S.; et al. Azithromycin, a 15-Membered Macrolide Antibiotic, Inhibits Influenza A(H1N1)Pdm09 Virus Infection by Interfering with Virus Internalization Process. *J. Antibiot. (Tokyo).* **2019**, *72*, 759–768. [CrossRef] [PubMed]
57. Retallack, H.; Di Lullo, E.; Arias, C.; Knopp, K.A.; Laurie, M.T.; Sandoval-Espinosa, C.; Mancia Leon, W.R.; Krencik, R.; Ullian, E.M.; Spatazza, J.; et al. Zika Virus Cell Tropism in the Developing Human Brain and Inhibition by Azithromycin. *Proc. Natl. Acad. Sci. USA* **2016**, *113*, 14408–14413. [CrossRef] [PubMed]
58. Madrid, P.B.; Panchal, R.G.; Warren, T.K.; Shurtleff, A.C.; Endsley, A.N.; Green, C.E.; Kolokoltsov, A.; Davey, R.; Manger, I.D.; Gilfillan, L.; et al. Evaluation of Ebola Virus Inhibitors for Drug Repurposing. *ACS Infect. Dis.* **2015**, *1*, 317–326. [CrossRef] [PubMed]
59. Damle, B.; Vourvahis, M.; Wang, E.; Leaney, J.; Corrigan, B. Clinical Pharmacology Perspectives on the Antiviral Activity of Azithromycin and Use in COVID-19. *Clin. Pharmacol. Ther.* **2020**, *108*, 201–211. [CrossRef]
60. Menzel, M.; Akbarshahi, H.; Bjermer, L.; Uller, L. Azithromycin Induces Anti-Viral Effects in Cultured Bronchial Epithelial Cells from COPD Patients. *Sci. Rep.* **2016**, *6*, 28698. [CrossRef]
61. Zeng, S.; Meng, X.; Huang, Q.; Lei, N.; Zeng, L.; Jiang, X.; Guo, X. Spiramycin and Azithromycin, Safe for Administration to Children, Exert Antiviral Activity against Enterovirus A71 in Vitro and in Vivo. *Int. J. Antimicrob. Agents* **2019**, *53*, 362–369. [CrossRef]
62. de Wit, E.; van Doremalen, N.; Falzarano, D.; Munster, V.J. SARS and MERS: Recent Insights into Emerging Coronaviruses. *Nat. Rev. Microbiol.* **2016**, *14*, 523–534. [CrossRef]
63. Ou, X.; Liu, Y.; Lei, X.; Li, P.; Mi, D.; Ren, L.; Guo, L.; Guo, R.; Chen, T.; Hu, J.; et al. Characterization of Spike Glycoprotein of SARS-CoV-2 on Virus Entry and Its Immune Cross-Reactivity with SARS-CoV. *Nat. Commun.* **2020**, *11*, 1620. [CrossRef] [PubMed]
64. Beigel, J.H.; Tomashek, K.M.; Dodd, L.E. Remdesivir for the Treatment of COVID-19 - Preliminary Report. Reply. *N. Engl. J. Med.* **2020**, *383*, 994. [CrossRef]
65. Wang, Y.; Zhang, D.; Du, G.; Du, R.; Zhao, J.; Jin, Y.; Fu, S.; Gao, L.; Cheng, Z.; Lu, Q.; et al. Remdesivir in Adults with Severe COVID-19: A Randomised, Double-Blind, Placebo-Controlled, Multicentre Trial. *Lancet (Lond. Engl.)* **2020**, *395*, 1569–1578. [CrossRef]
66. Williamson, B.N.; Feldmann, F.; Schwarz, B.; Meade-White, K.; Porter, D.P.; Schulz, J.; van Doremalen, N.; Leighton, I.; Yinda, C.K.; Pérez-Pérez, L.; et al. Clinical Benefit of Remdesivir in Rhesus Macaques Infected with SARS-CoV-2. *Nature* **2020**, *585*, 273–276. [CrossRef] [PubMed]
67. Andreani, J.; Le Bideau, M.; Duflot, I.; Jardot, P.; Rolland, C.; Boxberger, M.; Wurtz, N.; Rolain, J.-M.; Colson, P.; La Scola, B.; et al. In Vitro Testing of Combined Hydroxychloroquine and Azithromycin on SARS-CoV-2 Shows Synergistic Effect. *Microb. Pathog.* **2020**, *145*, 104228. [CrossRef]
68. Touret, F.; Gilles, M.; Barral, K.; Nougairède, A.; van Helden, J.; Decroly, E.; de Lamballerie, X.; Coutard, B. In Vitro Screening of a FDA Approved Chemical Library Reveals Potential Inhibitors of SARS-CoV-2 Replication. *Sci. Rep.* **2020**, *10*, 13093. [CrossRef]
69. Gautret, P.; Lagier, J.-C.; Parola, P.; Hoang, V.T.; Meddeb, L.; Mailhe, M.; Doudier, B.; Courjon, J.; Giordanengo, V.; Vieira, V.E.; et al. Hydroxychloroquine and Azithromycin as a Treatment of COVID-19: Results of an Open-Label Non-Randomized Clinical Trial. *Int. J. Antimicrob. Agents* **2020**, *56*, 105949. [CrossRef]
70. Gyselinck, I.; Janssens, W.; Verhamme, P.; Vos, R. Rationale for Azithromycin in COVID-19: An Overview of Existing Evidence. *BMJ open Respir. Res.* **2021**, *8*. [CrossRef]
71. Maisonnasse, P.; Guedj, J.; Contreras, V.; Behillil, S.; Solas, C.; Marlin, R.; Naninck, T.; Pizzorno, A.; Lemaitre, J.; Gonçalves, A.; et al. Hydroxychloroquine Use against SARS-CoV-2 Infection in Non-Human Primates. *Nature* **2020**, *585*, 584–587. [CrossRef]
72. Albani, F.; Fusina, F.; Giovannini, A.; Ferretti, P.; Granato, A.; Prezioso, C.; Divizia, D.; Sabaini, A.; Marri, M.; Malpetti, E.; et al. Impact of Azithromycin and/or Hydroxychloroquine on Hospital Mortality in COVID-19. *J. Clin. Med.* **2020**, *9*, 280. [CrossRef]
73. Arshad, S.; Kilgore, P.; Chaudhry, Z.S.; Jacobsen, G.; Wang, D.D.; Huitsing, K.; Brar, I.; Alangaden, G.J.; Ramesh, M.S.; McKinnon, J.E.; et al. Treatment with Hydroxychloroquine, Azithromycin, and Combination in Patients Hospitalized with COVID-19. *Int. J. Infect. Dis. IJID Off. Publ. Int. Soc. Infect. Dis.* **2020**, *97*, 396–403. [CrossRef] [PubMed]
74. Sekhavati, E.; Jafari, F.; SeyedAlinaghi, S.; Jamalimoghadamsiahkali, S.; Sadr, S.; Tabarestani, M.; Pirhayati, M.; Zendehdel, A.; Manafi, N.; Hajiabdolbaghi, M.; et al. Safety and Effectiveness of Azithromycin in Patients with COVID-19: An Open-Label Randomised Trial. *Int. J. Antimicrob. Agents* **2020**, *56*, 106143. [CrossRef] [PubMed]
75. Lauriola, M.; Pani, A.; Ippoliti, G.; Mortara, A.; Milighetti, S.; Mazen, M.; Perseghin, G.; Pastori, D.; Grosso, P.; Scaglione, F. Effect of Combination Therapy of Hydroxychloroquine and Azithromycin on Mortality in Patients with COVID-19. *Clin. Transl. Sci.* **2020**, *13*, 1071–1076. [CrossRef] [PubMed]
76. Tanriverdİ, E.; Çörtük, M.; Yildirim, B.Z.; Uğur Chousein, E.G.; Turan, D.; Çinarka, H.; Özgül, M.A.; Çetinkaya, E. Hydroxychloroquine plus Azithromycin and Early Hospital Admission Are Beneficial in COVID-19 Patients: Turkish Experience with Real-Life Data. *Turkish J. Med. Sci.* **2021**, *51*, 10–15. [CrossRef]
77. Guérin, V.; Lévy, P.; Thomas, J.-L.; Lardenois, T.; Lacrosse, P.; Sarrazin, E.; de Andreis, N.R.; Wonner, M. Azithromycin and Hydroxychloroquine Accelerate Recovery of Outpatients with Mild/Moderate COVID-19. *Asian J. Med. Heal.* **2020**, *18*, 15–55. [CrossRef]

78. Kuderer, N.M.; Choueiri, T.K.; Shah, D.P.; Shyr, Y.; Rubinstein, S.M.; Rivera, D.R.; Shete, S.; Hsu, C.-Y.; Desai, A.; de Lima Lopes, G.J.; et al. Clinical Impact of COVID-19 on Patients with Cancer (CCC19): A Cohort Study. *Lancet (Lond. Engl.)* **2020**, *395*, 1907–1918. [CrossRef]
79. Rosenberg, E.S.; Dufort, E.M.; Udo, T.; Wilberschied, L.A.; Kumar, J.; Tesoriero, J.; Weinberg, P.; Kirkwood, J.; Muse, A.; DeHovitz, J.; et al. Association of Treatment with Hydroxychloroquine or Azithromycin with In-Hospital Mortality in Patients with COVID-19 in New York State. *JAMA* **2020**, *323*, 2493–2502. [CrossRef]
80. Cavalcanti, A.B.; Zampieri, F.G.; Rosa, R.G.; Azevedo, L.C.P.; Veiga, V.C.; Avezum, A.; Damiani, L.P.; Marcadenti, A.; Kawano-Dourado, L.; Lisboa, T.; et al. Hydroxychloroquine with or without Azithromycin in Mild-to-Moderate COVID-19. *N. Engl. J. Med.* **2020**, *383*, 2041–2052. [CrossRef]
81. Ayerbe, L.; Risco-Risco, C.; Ayis, S. The Association of Treatment with Hydroxychloroquine and Hospital Mortality in COVID-19 Patients. *Intern. Emerg. Med.* **2020**, *15*, 1501–1506. [CrossRef]
82. Lammers, A.J.J.; Brohet, R.M.; Theunissen, R.E.P.; Koster, C.; Rood, R.; Verhagen, D.W.M.; Brinkman, K.; Hassing, R.J.; Dofferhoff, A.; El Moussaoui, R.; et al. Response to Correspondence Concerning: "Early Hydroxychloroquine but Not Chloroquine Use Reduces ICU Admission in COVID-19 Patients". *Int. J. Infect. Dis. IJID Off. Publ. Int. Soc. Infect. Dis.* **2021**, *103*, 478–479. [CrossRef]
83. Kamel, A.M.; Monem, M.S.A.; Sharaf, N.A.; Magdy, N.; Farid, S.F. Efficacy and Safety of Azithromycin in COVID-19 Patients: A Systematic Review and Meta-Analysis of Randomized Clinical Trials. *Rev. Med. Virol.* **2022**, *32*, e2258. [CrossRef] [PubMed]
84. Echeverría-Esnal, D.; Martin-Ontiyuelo, C.; Navarrete-Rouco, M.E.; De-Antonio Cuscó, M.; Ferrández, O.; Horcajada, J.P.; Grau, S. Azithromycin in the Treatment of COVID-19: A Review. *Expert Rev. Anti. Infect. Ther.* **2021**, *19*, 147–163. [CrossRef] [PubMed]
85. Szente Fonseca, S.N.; de Queiroz Sousa, A.; Wolkoff, A.G.; Moreira, M.S.; Pinto, B.C.; Valente Takeda, C.F.; Rebouças, E.; Vasconcellos Abdon, A.P.; Nascimento, A.L.A.; Risch, H.A. Risk of Hospitalization for COVID-19 Outpatients Treated with Various Drug Regimens in Brazil: Comparative Analysis. *Travel Med. Infect. Dis.* **2020**, *38*, 101906. [CrossRef]
86. Geleris, J.; Sun, Y.; Platt, J.; Zucker, J.; Baldwin, M.; Hripcsak, G.; Labella, A.; Manson, D.K.; Kubin, C.; Barr, R.G.; et al. Observational Study of Hydroxychloroquine in Hospitalized Patients with COVID-19. *N. Engl. J. Med.* **2020**, *382*, 2411–2418. [CrossRef] [PubMed]
87. Furtado, R.H.M.; Berwanger, O.; Fonseca, H.A.; Corrêa, T.D.; Ferraz, L.R.; Lapa, M.G.; Zampieri, F.G.; Veiga, V.C.; Azevedo, L.C.P.; Rosa, R.G.; et al. Azithromycin in Addition to Standard of Care versus Standard of Care Alone in the Treatment of Patients Admitted to the Hospital with Severe COVID-19 in Brazil (COALITION II): A Randomised Clinical Trial. *Lancet (Lond. Engl.)* **2020**, *396*, 959–967. [CrossRef]
88. Azithromycin for Community Treatment of Suspected COVID-19 in People at Increased Risk of an Adverse Clinical Course in the UK (PRINCIPLE): A Randomised, Controlled, Open-Label, Adaptive Platform Trial. *Lancet (Lond. Engl.)* **2021**, *397*, 1063–1074. [CrossRef]
89. Hinks, T.S.C.; Barber, V.S.; Black, J.; Dutton, S.J.; Jabeen, M.; Melhorn, J.; Rahman, N.M.; Richards, D.; Lasserson, D.; Pavord, I.D.; et al. A Multi-Centre Open-Label Two-Arm Randomised Superiority Clinical Trial of Azithromycin versus Usual Care in Ambulatory COVID-19: Study Protocol for the ATOMIC2 Trial. *Trials* **2020**, *21*, 718. [CrossRef]
90. Tocilizumab in Patients Admitted to Hospital with COVID-19 (RECOVERY): A Randomised, Controlled, Open-Label, Platform Trial. *Lancet (Lond. Engl.)* **2021**, *397*, 1637–1645. [CrossRef]
91. Bogdanić, N.; Močibob, L.; Vidović, T.; Soldo, A.; Begovać, J. Azithromycin Consumption during the COVID-19 Pandemic in Croatia, 2020. *PLoS One* **2022**, *17*, e0263437. [CrossRef]
92. Gouin, K.A.; Creasy, S.; Beckerson, M.; Wdowicki, M.; Hicks, L.A.; Lind, J.N.; Geller, A.I.; Budnitz, D.S.; Kabbani, S. Trends in Prescribing of Antibiotics and Drugs Investigated for Coronavirus Disease 2019 (COVID-19) Treatment in US Nursing Home Residents During the COVID-19 Pandemic. *Clin. Infect. Dis. an Off. Publ. Infect. Dis. Soc. Am.* **2022**, *74*, 74–82. [CrossRef]
93. Stoichitoiu, L.E.; Pinte, L.; Ceasovschih, A.; Cernat, R.C.; Vlad, N.D.; Padureanu, V.; Sorodoc, L.; Hristea, A.; Purcarea, A.; Badea, C.; et al. In-Hospital Antibiotic Use for COVID-19: Facts and Rationales Assessed through a Mixed-Methods Study. *J. Clin. Med.* **2022**, *11*, 3194. [CrossRef] [PubMed]
94. Prestinaci, F.; Pezzotti, P.; Pantosti, A. Antimicrobial Resistance: A Global Multifaceted Phenomenon. *Pathog. Glob. Health* **2015**, *109*, 309–318. [CrossRef] [PubMed]
95. Gomes, C.; Ruiz-Roldán, L.; Mateu, J.; Ochoa, T.J.; Ruiz, J. Azithromycin Resistance Levels and Mechanisms in Escherichia Coli. *Sci. Rep.* **2019**, *9*, 6089. [CrossRef] [PubMed]
96. Bommer, U.; Burkhardt, N.; Junemann, R.; Spahn, C.M.T.; Triana- Alonso, F.J.; Nierhaus, K.H. Ribosomes and Polysomes. In *Subcellular Fractionation. A Practical Approach*; Graham, J.D., Rickwoods, E., Eds.; IRL Press at Oxford University Press: Oxford, UK, 1996; pp. 271–301.
97. Murray, C.J.; Ikuta, K.S.; Sharara, F.; Swetschinski, L.; Aguilar, G.R.; Gray, A.; Han, C.; Bisignano, C.; Rao, P.; Wool, E.; et al. Global Burden of Bacterial Antimicrobial Resistance in 2019: A Systematic Analysis. *Lancet (Lond. Engl.)* **2022**, *399*, 629–655. [CrossRef]
98. Lansbury, L.; Lim, B.; Baskaran, V.; Lim, W.S. Co-Infections in People with COVID-19: A Systematic Review and Meta-Analysis. *J. Infect.* **2020**, *81*, 266–275. [CrossRef]
99. Chertow, D.S.; Memoli, M.J. Bacterial Coinfection in Influenza: A Grand Rounds Review. *JAMA* **2013**, *309*, 275–282. [CrossRef]

100. Ishaqui, A.A.; Khan, A.H.; Sulaiman, S.A.S.; Alsultan, M.T.; Khan, I.; Naqvi, A.A. Assessment of Efficacy of Oseltamivir Azithromycin Combination Therapy in Prevention of Influenza-A (H1N1)Pdm09 Infection Complications and Rapidity of Symptoms Relief. *Expert Rev. Respir. Med.* **2020**, *14*, 533–541. [CrossRef]
101. Lee, N.; Wong, C.-K.; Chan, M.C.W.; Yeung, E.S.L.; Tam, W.W.S.; Tsang, O.T.Y.; Choi, K.-W.; Chan, P.K.S.; Kwok, A.; Lui G.C.Y.; et al. Anti-Inflammatory Effects of Adjunctive Macrolide Treatment in Adults Hospitalized with Influenza: A Randomized Controlled Trial. *Antiviral Res.* **2017**, *144*, 48–56. [CrossRef]
102. Segala, F.V.; Bavaro, D.F.; Di Gennaro, F.; Salvati, F.; Marotta, C.; Saracino, A.; Murri, R.; Fantoni, M. Impact of SARS-CoV-2 Epidemic on Antimicrobial Resistance: A Literature Review. *Viruses* **2021**, *13*, 2110. [CrossRef]
103. Pinte, L.; Ceasovschih, A.; Niculae, C.-M.; Stoichitoiu, L.E.; Ionescu, R.A.; Balea, M.I.; Cernat, R.C.; Vlad, N.; Padureanu, V. Purcarea, A.; et al. Antibiotic Prescription and In-Hospital Mortality in COVID-19: A Prospective Multicentre Cohort Study. *J Pers. Med.* **2022**, *12*, 0877. [CrossRef]

antibiotics

MDPI

Article

Co-Administration of Remdesivir and Azithromycin May Protect against Intensive Care Unit Admission in COVID-19 Pneumonia Requiring Hospitalization: A Real-Life Observational Study

Andrea Ticinesi [1,2,*], Domenico Tuttolomondo [1,3], Antonio Nouvenne [2], Alberto Parise [2], Nicoletta Cerundolo [2], Beatrice Prati [2], Ilaria Zanichelli [1], Angela Guerra [1,2], Nicola Gaibazzi [3] and Tiziana Meschi [1,2]

1 Department of Medicine and Surgery, University of Parma, Via Antonio Gramsci 14, 43126 Parma, Italy; domenico.tuttolomondo@unipr.it (D.T.); ilaria.zanichelli@unipr.it (I.Z.); angela.guerra@unipr.it (A.G.); tiziana.meschi@unipr.it (T.M.)

2 Geriatric-Rehabilitation Department, Azienda Ospedaliero-Universitaria di Parma, Via Antonio Gramsci 14, 43126 Parma, Italy; anouvenne@ao.pr.it (A.N.); aparise@ao.pr.it (A.P.); ncerundolo@ao.pr.it (N.C.); bprati@ao.pr.it (B.P.)

3 Department of Cardiology, Azienda Ospedaliero-Universitaria di Parma, Via Antonio Gramsci 14, 43126 Parma, Italy; ngaibazzi@ao.pr.it

* Correspondence: andrea.ticinesi@unipr.it

Citation: Ticinesi, A.; Tuttolomondo, D.; Nouvenne, A.; Parise, A.; Cerundolo, N.; Prati, B.; Zanichelli, I.; Guerra, A.; Gaibazzi, N.; Meschi, T. Co-Administration of Remdesivir and Azithromycin May Protect against Intensive Care Unit Admission in COVID-19 Pneumonia Requiring Hospitalization: A Real-Life Observational Study. *Antibiotics* **2022**, *11*, 941. https://doi.org/10.3390/antibiotics11070941

Academic Editors: Johan Van Laethem and Sabine Danielle Allard

Received: 24 June 2022
Accepted: 12 July 2022
Published: 14 July 2022

Abstract: The benefits of remdesivir treatment, with or without co-administration of antibiotics such as azithromycin, are uncertain in COVID-19 pneumonia. The aim of this retrospective single-center study was to assess the effects of remdesivir, with or without azithromycin, on hospital mortality, intensive care unit (ICU) admission, and need of non-invasive ventilation. The clinical records of the COVID-19 patients hospitalized in an Italian ward in March 2021 were analyzed, and data on comorbidities and clinical, radiological, and laboratory presentation of the disease were collected. Among 394 participants (234 M), 173 received remdesivir (43.9%), including 81 with azithromycin (20.5%). Remdesivir recipients were younger, with less comorbidities, and had better PaO_2/FiO_2 and clinical outcomes, including reduced mortality, but the differences were not independent of covariates. Rates of ICU transferal were 17%, 9%, and 1% in the no remdesivir, remdesivir without azithromycin, and remdesivir/azithromycin groups, respectively. In a stepwise multivariate logistic regression model, remdesivir/azithromycin co-treatment was independently associated with reduced ICU admission (vs remdesivir alone, OR 0.081, 95% CI 0.008–0.789, $p = 0.031$; vs no remdesivir, OR 0.060, 95% CI 0.007–0.508, $p = 0.010$). These data suggest that the therapeutical effect of remdesivir in COVID-19 pneumonia may be potentiated by azithromycin. The association between the two drugs should be further investigated.

Keywords: SARS-CoV-2; antiviral; antibiotic; azithromycin; remdesivir

1. Introduction

Current guidelines consider intravenous steroid treatment as the mainstay of pharmacological treatment of severe and critical forms of COVID-19 pneumonia, provided that it is not administered in the earliest days of symptoms [1,2]. Steroids, and particularly dexamethasone, have in fact proved effective in reducing mortality and, to a lesser extent, also the duration of mechanical ventilation, if needed [1–3]. In selected cases, steroids can be administered in association with IL-6 receptor blockers or baricitinib [1,2].

Antiviral drugs available before emergence of the pandemic have shown little or no efficacy in reducing mortality or mechanical ventilation in severe forms of COVID-19 requiring hospitalization [1,2]. The only exception is represented by remdesivir, whose administration within the first ten days of symptoms to patients with non-critical illness has shown some benefits in terms of need of ventilation and its duration, but no effect on mortality, in comparison with support treatment only [1,2,4–6].

However, real-life studies investigating the clinical outcomes of patients receiving intravenous remdesivir treatment during hospitalization for severe COVID-19 pneumonia have produced conflicting results. Some reports have shown little or no impact from this treatment, either alone or in association with steroids, on clinically relevant outcomes such as intensive care unit (ICU) admission or mortality [7–10]. Conversely, other studies suggested that it could be useful, as an add-on treatment, in adults with persistent fever and dyspnea and moderate lung involvement on chest CT, reducing ICU admission, length of hospital stay, and even mortality [11–15], especially when administered in the first 5 days of symptoms [15].

In the earliest phases of the COVID-19 pandemic, the use of azithromycin has also been proposed as a therapeutical option for its potential antiviral and anti-inflammatory effects shown in vitro [16]. Trials investigating the efficacy of azithromycin alone against severe forms of COVID-19 have produced negative results, and its use is no longer recommended [1,2,17,18]. However, antibiotic treatment is still part of the clinical management of severe COVID-19, especially in patients showing multiple pulmonary infiltrates on chest imaging or marked elevation of serum inflammation indexes suggesting the presence of a high risk of bacterial superinfection [2,19]. In these cases, azithromycin is commonly used for its antibacterial action with low risk of generating resistance.

To date, there is still uncertainty on the clinical effectiveness of the association between remdesivir and other drugs showing in vitro activity against SARS-CoV-2, including azithromycin, in patients with severe COVID-19 pneumonia requiring hospitalization. Furthermore, the real-life clinical benefits of remdesivir are far from fully understood, especially in older patients with multiple comorbidities, who are typically excluded from clinical trials.

The aims of this real-life observational study were to describe the clinical characteristics of patients receiving remdesivir during the third pandemic wave (March 2021) in an Italian hub for the care of COVID-19, and to assess the effects of remdesivir treatment, either with or without azithromycin, on clinical outcomes.

2. Materials and Methods

2.1. Study Setting, Design, and Inclusion Criteria

This was a retrospective, real-life, observational study conducted in the context of the third pandemic wave in Italy in March 2021, which was mainly sustained by the emergence of the alpha SARS-CoV-2 variant [20]. The setting of the study was the Internal Medicine Unit of the Geriatric-Rehabilitation Department of Parma University-Hospital, which had been identified as the main hub for the hospital care of patients with COVID-19 of the whole Parma Province (>450,000 inhabitants) since the first wave in 2020 [21,22].

Included in the study were all patients admitted 1–31 March 2021 with positive nasopharyngeal swab for SARS-CoV-2, presence of symptoms compatible with COVID-19, and chest Computed Tomography (CT) results that were positive or indeterminate for the presence of COVID-19 pneumonia. Excluded were patients who did not undergo chest imaging, patients with missing data on treatments against COVID-19 administered before and during hospital stay, and patients who explicitly denied their consent for study inclusion and data treatment.

2.2. Procedures and Data Collection

The clinical records and discharge forms of all eligible patients were reviewed by trained staff members. Data on the clinical presentation of COVID-19 (type and duration of symptoms, vital signs, chest CT findings, routine lab tests, and arterial blood gas analysis), number and type of comorbidities, and number of chronic medications were collected on case report forms. Namely, the extension of lung abnormalities on CT was measured through visual scoring, expressed as percentage of the lung parenchyma with ground-glass pattern or consolidations [23]. Respiratory function was assessed through calculation of the ratio between arterial oxygen partial pressure on arterial blood gas analysis and fraction of inspired oxygen on admission (PaO_2/FiO_2). Lab tests included serum levels of the main inflammatory markers upon admission: C-reactive protein (CRP), procalcitonin (PCT), and IL-6. The severity of COVID-19 was also classified according to the WHO COVID-19 Scale [24,25].

Treatments administered against COVID-19 (steroids, IL-6 receptor antagonists, remdesivir, antibiotics including azithromycin) were also collected by review of clinical records, along with the maximal oxygen or ventilatory need during hospital stay. The hospitalization outcome (discharge vs death) was also assessed.

2.3. Exposure and Outcome Variables

Patients included in the study were treated in accordance with the recommendations for COVID-19 in force in Italy in March 2021 [26]. Allocation to treatments (corticosteroids, enoxaparin, IL-6 antagonists, remdesivir, antibiotics including azithromycin) was independent of the study. Treatment with remdesivir, either in association with azithromycin or not, was considered the main exposure variable for analyses.

The decision to treat patients with remdesivir was based on the compliance with criteria of prescription issued by the European Medicines Agency and by the Italian Drug Agency (AIFA—Agenzia Italiana del Farmaco) [27] and on the clinical judgement of the prescribing physician. According to AIFA criteria, remdesivir could be prescribed to hospitalized patients with COVID-19 only in case all of the following were satisfied:

- duration of COVID-19 related symptoms not exceeding 10 days.
- presence of lung parenchyma abnormalities on chest imaging.
- no need of oxygen therapy with high-flow nasal cannulae (HFNC), non-invasive or invasive mechanical ventilation, or extracorporeal membrane oxygenation (ECMO).
- serum aspartate and alanine aminotransferase levels not exceeding 5 times the upper limit of reference.
- glomerular filtration rate (GFR) not inferior to 30 mL/min.

Remdesivir was administered intravenously with a load dose of 200 mg on day 1, followed by maintenance doses of 100 mg daily from day 2 to day 5.

The decision to treat patients with azithromycin relied on clinical judgement and was based on the risk of bacterial superinfection of COVID-19-related pulmonary lesions after review of the clinical presentation, chest imaging, and lab tests, including CRP and PCT levels. Azithromycin was administered orally at 500 mg per day for three to five days, according to clinical judgement.

Hospital mortality was considered the main study endpoint. ICU admission, need of invasive or non-invasive mechanical ventilation, and length of overall hospital stay were also considered as secondary endpoints.

2.4. Statistical Analyses

Data were expressed as percentages or median and interquartile range (IQR), as appropriate. The clinical characteristics and outcomes were compared after categorizing participants for treatment with remdesivir, and with remdesivir plus azithromycin. When comparing continuous variables, the Mann–Whitney and Kruskal–Wallis tests were used. Significant values were adjusted for potential confounders with the Quade non-parametric Ancova test. Chi-square test and binary logistic regression were used for comparing discrete variables.

Stepwise logistic regression models were used to verify the associations between exposures (particularly treatment with remdesivir and remdesivir/azithromycin association) with primary or secondary outcomes. Age, sex, chest CT visual score, PaO_2/FiO_2 on admission, and the number of chronic comorbidities were considered as covariates.

Statistical analyses were performed with the SPSS package (v.28, IBM, Armonk, NY USA); p values were considered significant when <0.05.

3. Results

3.1. General Characteristics of the Population

In March 2021, 467 patients with COVID-19 were admitted to the study unit. After checking for inclusion and exclusion criteria, 394 patients (234 M, 160 F) were finally included. Among them, 173 patients (43.9%) received remdesivir during hospital stay. Azithromycin was co-administered in 81 remdesivir recipients (20.5% of the whole population) (Supplementary Figure S1).

A comparison between the baseline clinical characteristics of patients who received remdesivir and those who did not is shown in Table 1. Namely, remdesivir recipients were younger (age median 60, IQR 52–71, vs. 71, IQR 61–80 years old, $p < 0.001$), had a reduced burden of comorbidities (median number of chronic illnesses 1, IQR 0–3, vs. 3, IQR 1–4, $p < 0.001$), and better respiratory exchanges (PaO_2/FiO_2 median 312, IQR 264–352, vs. 276, IQR 202–328 mmHg, $p < 0.001$). Serum inflammatory markers and neutrophil count were also lower in remdesivir recipients upon hospital admission.

Table 1. Comparison of the main characteristics of patients treated with remdesivir and patients who did not receive the drug.

	No Remdesivir (N = 221)	Treated with Remdesivir (N = 173)	*p*
	Demography and personal history		
Age, years	71 (61–80)	60 (52–71)	**<0.001**
Female sex, %	42	39	0.641
Chronic comorbidities, number	3 (1–4)	1 (0–3)	**<0.001**
Chronic medications, number	2 (1–5)	1 (0–3)	**<0.001**
Hypertension, %	59	45	**0.005**
Ischemic heart disease, %	15	5	**0.001**
Diabetes, %	21	13	**0.040**
Obesity, %	13	17	0.194
Dyslipidemia, %	21	17	0.309
Atrial Fibrillation, %	16	6	**0.001**
	Chest CT presentation		
Ground glass abnormalities on CT, %	90	99	**<0.001**
Consolidations on CT, %	69	71	0.782
Visual score of pneumonia extension, %	25 (15–45)	20 (15–35)	0.074
	Clinical presentation upon admission		
PaO_2/FiO_2, mmHg	276 (202–328)	312 (264–352)	**<0.001**
Duration of symptoms, days	6 (2–10)	7 (4–9)	0.150
Fever, %	76	81	0.286
Cough, %	44	53	0.082
Dyspnea,%	58	43	**0.003**
Diarrhoea, %	17	16	0.790
WHO COVID-19 Scale mild, %	10	1	**<0.001**
WHO COVID-19 Scale moderate, %	24	32	0.082
WHO COVID-19 Scale severe, %	27	42	**0.002**
WHO COVID-19 Scale critical, %	38	24	**0.003**

Table 1. *Cont.*

	No Remdesivir (N = 221)	Treated with Remdesivir (N = 173)	*p*
	Blood tests		
Hemoglobin, g/dL	13.8 (12.2–15.0)	14.0 (12.9–15.0)	0.055
Platelet count, 1000/mm^3	195 (146–266)	187 (149–223)	0.125
White Blood Cell count, n/mm^3	6980 (5113–8885)	5500 (4060–7850)	**<0.001**
Neutrophil count, n/mm^3	5403 (3610–7392)	4400 (2781–6243)	**<0.001**
Creatinine, mg/dL	0.9 (0.7–1.2)	0.8 (0.7–1.0)	**0.004**
C-Reactive Protein, mg/L	62 (29–107)	46 (23–81)	**0.005**
Procalcitonin, ng/mL	0.11 (0.06–0.39)	0.07 (0.04–0.14)	**<0.001**
Interleukin-6, pg/mL	86 (29–182)	79 (34–145)	0.908
D-dimer, ng/mL	940 (532–1635)	608 (420–906)	**<0.001**
	Other treatments administered		
Enoxaparin, %	95	97	0.401
Corticosteroids,%	90	99	**<0.001**
Antibiotics, %	88	87	0.750
Azithromycin, %	29	47	**<0.001**

Data are shown as median and interquartile range or percentages. Crude comparisons were made with Mann–Whitney test or chi-square test, as appropriate; *p* values < 0.05 are indicated in bold.

Table 2 shows multiple comparisons of baseline clinical characteristics, after splitting the group of remdesivir recipients between those who received remdesivir alone and those who received remdesivir plus azithromycin. These two groups were comparable for demographic, clinical, and laboratory characteristics.

Table 2. Comparison of the main characteristics of patients who were not treated with remdesivir (1), who received remdesivir but not azithromycin (2), and who received remdesivir plus azithromycin (3).

	No Remdesivir N = 221 (1)	Treated with Remdesivir, No Azithromycin N = 92 (2)	Treated with Remdesivir and Azithromycin N = 81 (3)	*p*	
		Demography and personal history			
Age, years	71 (61–80)	59 (52–70)	60 (51–72)	**<0.001**	(1) vs. (2) vs. (3)
Female sex, %	42	36	43	0.555	
Chronic comorbidities, number	3 (1–4)	1 (1–3)	1 (0–3)	**<0.001**	(1) vs. (2) vs. (3)
Chronic medications, number	2 (1–5)	1 (0–3)	1 (0–2)	**<0.001**	(1) vs. (2) vs. (3)
Hypertension, %	59	49	41	**0.012**	(1) vs. (3)
Ischemic heart disease, %	15	7	4	**0.009**	(1) vs. (2) vs. (3)
Diabetes, %	21	15	11	0.100	
Obesity, %	13	20	15	0.297	
Dyslipidemia, %	21	15	19	0.515	
Atrial fibrillation, %	16	8	4	**0.008**	(1) vs. (2) vs. (3)
		Chest CT presentation			
Ground glass abnormalities on CT, %	90	99	99	**0.011**	(1) vs. (2) vs. (3)
Consolidations on CT, %	69	70	72	0.922	
Visual score of pneumonia extension, %	25 (15–45)	20 (15–35)	20 (15–30)	0.153	

Table 2. *Cont.*

	No Remdesivir N = 221 (1)	Treated with Remdesivir, No Azithromycin N = 92 (2)	Treated with Remdesivir and Azithromycin N = 81 (3)	*p*	
		Clinical presentation upon admission			
PaO_2/FiO_2, mmHg	276 (202–328)	319 (273–354)	305 (262–348)	**<0.001**	(1) vs. (2) vs. (3)
Duration of symptoms, days	6 (2–10)	7 (4–9)	6 (4–9)	0.352	
Fever, %	76	80	81	0.559	
Cough, %	44	55	51	0.181	
Dyspnea,%	58	52	32	**<0.001**	(3) vs. (1) vs. (2)
Anosmia, %	7	8	14	0.171	
Diarrhoea, %	17	18	14	0.668	
WHO COVID-19 Scale mild, %	10	1	1	**0.011**	(1) vs. (2) vs. (3)
WHO COVID-19 Scale moderate, %	24	30	35	0.185	
WHO COVID-19 Scale severe, %	27	46	38	**0.005**	(2) vs. (1)
WHO COVID-19 Scale critical, %	38	23	26	**0.011**	(1) vs. (2) vs. (3)
		Blood tests			
Hemoglobin, g/dL	13.8 (12.2–15.0)	14.0 (12.9–14.9)	14.0 (12.9–15.0)	0.155	
Platelet count, 1000/mm^3	195 (146–266)	189 (152–224)	185 (148–222)	0.302	
White Blood Cell count, n/mm^3	6980 (5113–8885)	5930 (4380–8620)	5415 (3758–7318)	**<0.001**	(1) vs. (2) vs. (3)
Neutrophil count, n/mm^3	5403 (3610–7392)	4574 (3223–6368)	3915 (2526–5822)	**<0.001**	(1) vs. (2) vs. (3)
Creatinine, mg/dL	0.9 (0.7–1.2)	0.9 (0.7–1.0)	0.8 (0.7–1.0)	**0.009**	(1) vs. (3)
C-reactive protein, mg/L	62 (29–107)	47 (23–81)	45 (24–86)	**0.018**	
Procalcitonin, ng/mL	0.11 (0.06–0.39)	0.06 (0.04–0.13)	0.08 (0.05–0.14)	**<0.001**	(1) vs. (2) vs. (3)
Interleukin-6, pg/mL	86 (29–182)	81 (32–155)	73 (41–140)	0.931	
D-dimer, ng/mL	940 (532–1635)	577 (381–909)	628 (441–895)	**<0.001**	(1) vs. (2) vs. (3)
		Other treatments administered			
Enoxaparin, %	95	98	96	0.623	
Corticosteroids,%	90	99	99	**0.011**	(1) vs. (2) vs. (3)
Antibiotics, %	88	75	100	**<0.001**	(1) vs. (2) vs. (3); (2) vs. (3)

Data are shown as median and interquartile range or percentages. *p* calculated with Kruskal–Wallis test (significance values adapted on Bonferroni correction for multiple testing) or binary logistic regression; *p* values < 0.05 are indicated in bold.

3.2. Effects of Treatments on Mortality

Eighty patients (20.3%) died during hospital stay, 64 (29%) in the group not receiving remdesivir and 16 (9%) in the remdesivir recipients. This difference was not statistically significant after adjustment for age, sex, and parameters of disease severity (Table 3). In the remdesivir group, mortality was not different between those who received azithromycin and those who did not (Table 4).

Table 3. Comparison of study outcomes between patients who received remdesivir during hospital stay and those who did not.

	No Remdesivir (N = 221)	Treated with Remdesivir (N = 173)	p (Unadjusted)	p Adjusted for Age, Sex	p Adjusted (Model 1)	p Adjusted (Model 2)	p Adjusted (Model 3)	p Adjusted (Model 4)
Hospital death, %	29	9	**<0.001**	**0.005**	**0.016**	0.054	0.409	0.555
NIV, %	34	18	**<0.001**	**<0.001**	**0.001**	0.054	0.095	0.290
ICU admission, %	17	5	**<0.001**	**<0.001**	**<0.001**	**0.004**	**0.011**	**0.026**
Invasive ventilation, %	7	2	**0.027**	**0.019**	**0.022**	0.071	0.176	0.191
Length of stay, days	14 (10–24)	12 (8–19)	**0.027**	0.197	0.336	0.795	0.294	0.417

Model 1 adjusted for age, sex, and number of chronic illnesses. Model 2 adjusted for age, sex, number of chronic illnesses, and chest CT visual score. Model 3 adjusted for age, sex, number of chronic illnesses, and PaO_2/FiO_2 on admission. Model 4 adjusted for age, sex, number of chronic illnesses, chest CT visual score, and PaO_2/FiO_2 on admission. Value for p calculated with binary logistic regression, except for length of stay (unadjusted values calculated with Mann–Whitney test, adjusted values calculated with Quade nonparametric Ancova test; p values < 0.05 are indicated in bold. NIV = Non-Invasive Ventilation; ICU = Intensive Care Unit.

Table 4. Comparison of study outcomes between patients who received remdesivir azithromycin during hospital stay and those who did not receive remdesivir.

	No Remdesivir (N = 221)	Remdesivir Plus Azithromycin (N = 81)	p (Unadjusted)	p Adjusted for Age, Sex	p Adjusted (Model 1)	p Adjusted (Model 2)	p Adjusted (Model 3)	p Adjusted (Model 4)
Hospital death, %	29	9	**<0.001**	**0.026**	0.070	0.223	0.232	0.445
NIV, %	34	15	**0.001**	**0.001**	**0.002**	**0.031**	**0.028**	0.117
ICU admission, %	17	1	**<0.001**	**0.003**	**0.003**	**0.009**	**0.006**	**0.009**
Length of stay, days	14 (10–24)	12 (9–19)	**0.043**	0.606	0.645	0.869	0.902	0.948

Model 1 adjusted for age, sex, and number of chronic illnesses. Model 2 adjusted for age, sex, number of chronic illnesses, and chest CT visual score. Model 3 adjusted for age, sex, number of chronic illnesses, and PaO_2/FiO_2 on admission. Model 4 adjusted for age, sex, number of chronic illnesses, chest CT visual score, and PaO_2/FiO_2 on admission. Value of p calculated with binary logistic regression, except for length of stay (unadjusted values calculated with Mann–Whitney test, adjusted values calculated with Quade nonparametric Ancova test; p values < 0.05 are indicated in bold. NIV = Non-Invasive Ventilation; ICU = Intensive Care Unit.

In a stepwise multivariable logistic regression model, the only factors significantly associated with mortality were age (OR 1.109, 95% CI 1.069–1.150, $p < 0.001$), number of chronic illnesses (OR 1.457, 95% CI 1.210–1.755, $p < 0.001$), chest CT visual score of lung parenchyma involvement (OR 1.027, 95% CI 1.009–1.046, $p = 0.004$), and PaO_2/FiO_2 on admission (OR 0.991, 95% CI 0.987–0.995, $p < 0.001$). Neither remdesivir nor azithromycin treatment were independently associated with mortality.

3.3. *Effects of Treatments on Other Clinical Outcomes*

Patients who received remdesivir during hospital stay experienced lower frequency of non-invasive mechanical ventilation (18% vs. 34%) and ICU admission (5% vs. 17%). They also had shorter duration of hospitalization (median 12, IQR 8–19, vs. 14, IQR 10–24 days). However, the differences were not statistically significant after adjustment for age, sex, and parameters of disease severity, except for ICU admission (Table 3). Similar results were obtained when comparing patients receiving remdesivir plus azithromycin and those not on remdesivir treatment (Table 4). The rate of ICU admission was significantly lower, and very low in absolute terms, in those who were treated with remdesivir plus azithromycin (1% vs. 9%, $p < 0.001$).

In a stepwise logistic regression model, accounting also for age, sex, number of chronic illnesses, chest CT visual score of lung parenchyma involvement, and PaO_2/FiO_2 on admission, the association of remdesivir plus azithromycin, but not remdesivir alone, was protective against ICU admission (Table 5).

Table 5. Factors independently associated with ICU admission in the studied population on a stepwise logistic regression model.

	Odds Ratio	95% Confidence Interval	p
Age, years	0.961	0.933–0.989	**0.007**
Female sex (vs. male)	0.426	0.182–0.977	**0.049**
PaO_2/FiO_2, mmHg	0.982	0.977–0.987	**<0.001**
Treatments			**0.034**
Association between remdesivir and azithromycin (vs. no remdesivir)	0.060	0.007–0.508	**0.010**
Association between remdesivir and azithromycin (vs. remdesivir alone)	0.081	0.008–0.789	**0.031**
Remdesivir (vs. no remdesivir)	0.743	0.273–2.023	0.560

Model also accounting for age, sex, number of chronic illnesses, chest CT visual score of lung parenchyma involvement, and PaO_2/FiO_2 on admission; p values < 0.05 are indicated in bold.

3.4. *Safety Issues*

Treatment with remdesivir was stopped in only three patients, due to onset of bradycardia in two cases and marked aminotransferase elevation in one case. No serious adverse events were reported in clinical records of all other remdesivir recipients included in the study.

4. Discussion

In a group of patients hospitalized with severe COVID-19 during the third pandemic wave in Italy, treatment with remdesivir, administered in compliance with the recommendations from regulatory authorities, was associated with reduced mortality and reduced need of escalating ventilatory support. However, these outcomes were not independent of the baseline clinical presentation of COVID-19 and demographical characteristic, which were substantially different between remdesivir recipients and other patients. The co-administration of remdesivir and azithromycin, instead, seemed associated with reduced need of ICU admission independently of covariates.

To the best of our knowledge, this is the first study specifically investigating the effect of the co-administration of remdesivir with azithromycin in the scientific literature on COVID-19.

Treatment with remdesivir emerged as a promising therapeutical option for COVID-19 in the second half of 2020, with evidence that it could reduce the recovery time and length of hospital stay, though not affecting mortality [28]. Systematic reviews and meta-analyses conducted on further randomized controlled trials substantially confirmed that, in patients without illness requiring ventilatory support, the administration of remdesivir is associated with some clinical benefit, with reduced need of escalating ventilatory support, but has no apparent effect on mortality [4–6]. The severity of clinical presentation of COVID-19 patients during the second and third pandemic wave in 2020–2021, before the completion of mass vaccination campaigns, and the absence of specific and effective drugs against SARS-CoV-2, have made remdesivir administration particularly frequent in the hospital setting, despite the absence of clear advantages against mortality. To date, real-life studies have provided conflicting results, with some reports suggesting clear benefits from remdesivir administration, and other yielding negative results [7–15].

Timing of remdesivir administration may be a critical point in this regard. When administered within five days from the symptom onset, remdesivir may be associated with reduced rate of clinical progression towards severe respiratory failure, in comparison with administration from day 5 to day 10 of symptom onset [29,30]. A recent randomized controlled trial suggests that precocious use of remdesivir in the pre-hospital setting is associated with dramatically lower rates of hospitalization and progression towards severe COVID-19 forms [31]. However, this drug requires intravenous administration, which is not always feasible in the community setting, especially in the context of a pandemic.

Notably, in our study, the outcomes of remdesivir were unrelated with the timing of its administration.

Few reports have evaluated possible favorable interactions between remdesivir and other drugs used in patients with COVID-19. However, in clinical practice, pharmacologic treatment of SARS-CoV-2 pneumonia is rarely based on a single drug, in contrast with the design of most clinical trials. Most of the patients included in our investigation received also corticosteroids and antibiotics, in addition to remdesivir. This circumstance could contribute to explain why remdesivir administration was not independently associated with clinical benefits, because its effect was probably masked by steroid treatment. On the other side, the reduced need of ICU admission in patients who were treated with both remdesivir and azithromycin suggests that the clinical benefits of remdesivir could be reinforced by the administration of other drugs exhibiting anti-SARS-CoV-2 activity in vitro, like azithromycin. Also, it may contribute to explain why real-life studies on remdesivir efficacy have produced conflicting results [7–15].

Azithromycin has known antiviral effects in vitro that have been demonstrated also against SARS-CoV-2 [32]. Its efficacy as stand-alone treatment against COVID-19 is negligible [17,18], and in the early phases of the pandemic, it was studied as a therapeutic option in association with hydroxychloroquine, with a paradoxical increase of mortality [33]. Our findings, instead, suggest that the interaction between remdesivir and azithromycin should be further studied, both at the molecular and clinical level, in order to optimize recommendations for the hospital management of severe COVID-19.

Our study also highlighted that, in clinical practice, remdesivir was mainly administered to patients below the age of 70, with low comorbidity burden, moderate lung abnormality extension on chest imaging, and fairly good respiratory involvement on hospital admission. These patients may have lower risk of adverse outcomes, independently of treatments administered during hospital stay. In fact, the dramatic differences in outcomes between remdesivir recipients and other patients disappeared after correction for age and parameters associated with disease severity on admission, such as the PaO_2/FiO_2 ratio. These data reinforce the importance of evaluating the PaO_2/FiO_2 ratio on admission of patients with COVID-19, because it may represent the single parameter with the highest outcome prediction capacity [34].

However, the majority of severe cases of COVID-19 generally involve older subjects, with several comorbidities and relevant pneumonia extension on chest imaging [22,34,35]. Age-related frailty represents a consistent risk factor for adverse outcomes in COVID-19 [35,36], also for the increased risk of bacterial superinfection [37]. Many older subjects included in our study were excluded from remdesivir treatment because their clinical presentation on admission was not compatible with the criteria of prescription [27]. The role of frailty as modifier of the efficacy of anti-COVID-19 treatments, including remdesivir, should be studied in the future.

Our study has some limitations that should be carefully considered in result interpretation. First, the retrospective design is not ideal for assessing outcomes of pharmacologic treatments, and selection bias could not be excluded. In fact, the presence of strict regulatory criteria for prescription of remdesivir in Italy [27], which were followed in patients included in the study, may have dramatically affected allocation to remdesivir treatment. This is the main reason why demographic and clinical differences between participants who were treated with remdesivir and those who were not were relevant. Thus, statistical correction for the main parameters may not have adequately accounted for these differences, making the two groups not fully comparable. Similarly, in patients who required antibiotic therapy, azithromycin was prescribed mainly based on clinical judgement and in case of moderate respiratory involvement, while patients with severe or critical illness received other broad-spectrum antibiotics. This circumstance may have also affected allocation to azithromycin and should be regarded as a possible source of bias.

Furthermore, the sample size was not sufficiently large to explore clinical benefits of remdesivir treatment on secondary outcomes. The study data also refer to a period of predominance of the SARS-CoV-2 alpha variant, preceding mass vaccination against SARS-CoV-2. Vaccination campaigns and the emergence of novel SARS-CoV-2 variants with increased transmission and different clinical presentation have substantially modified the characteristics of patients hospitalized with COVID-19 [38], making the results of the present investigation not automatically transferrable to the current epidemiological situation.

In spite of these limitations, our results could be preliminary to the design of studies investigating the effects of the combination of remdesivir treatment with other drugs on the clinical course of COVID-19 pneumonia, and could serve as hypothesis generators for a better comprehension of the in vivo effects of remdesivir.

5. Conclusions

In a real-life setting during the third pandemic wave in Italy, the administration of remdesivir was not independently associated with improved mortality and reduced need of ventilatory support in patients hospitalized with severe COVID-19. However, the co-administration of remdesivir and azithromycin was associated with reduced risk of ICU admission, independently of covariates. The clinical efficacy of the association between remdesivir and azithromycin in COVID-19 should be further investigated.

Supplementary Materials: The following supporting information can be downloaded at: https://www.mdpi.com/article/10.3390/antibiotics11070941/s1, Supplementary Figure S1–Modified CONSORT diagram of the study.

Author Contributions: Conceptualization and methodology: A.T., A.N., N.G. and T.M.; investigation: A.T., D.T., A.P., N.C., B.P. and I.Z.; formal analysis: A.T. and A.G.; manuscript draft: A.T.; supervision: N.G. and T.M.; manuscript revision for substantial content: all authors. All authors have read and agreed to the published version of the manuscript.

Funding: Given the retrospective design of the study, no specific funding must be reported.

Institutional Review Board Statement: The study was approved by the local Ethics Committee (Comitato Etico dell'Area Vasta Emilia Nord, Emilia-Romagna Region, ID 399/2021/OSS/AOUPR, date of approval 9 June 2021) as part of a larger retrospective project on clinical and radiological factors associated with mortality in hospitalized COVID-19 patients. The study was conducted in compliance with Good Clinical Practice, the Declaration of Helsinki and its later amendments.

Informed Consent Statement: Informed consent was collected in written form from all patients who were contactable by the research team during the study period. For all other patients (either deceased or uncontactable upon reasonable effort), informed consent collection was waived.

Data Availability Statement: Access to data, in anonymous form, can be obtained upon reasonable and motivated request to the corresponding author. The subject entitled for data control and management is the Parma University-Hospital (Azienda Ospedaliero-Universitaria di Parma).

Acknowledgments: The authors wish to thank Maria Nicastro (University of Parma) for support in patient selection, and Franco Lori (Virostatics Srl, Sassari, Italy) for advice on study design and methodology.

Conflicts of Interest: The authors declare no conflict of interest.

References

1. Agarwal, A.; Rochwerg, B.; Lamontagne, F.; Siemieniuc, R.C.; Agoritsas, T.; Askie, L.; Lytvyn, L.; Leo, Y.S.; MacDonald, H.; Zeng, L.; et al. A living WHO guideline on drugs for COVID-19. *BMJ* **2020**, *370*, m3379. [CrossRef] [PubMed]
2. Bartoletti, M.; Azap, O.; Barac, A.; Bussini, L.; Ergonul, O.; Krause, R.; Paño-Pardo, J.R.; Power, N.R.; Sibani, M.; Szabo, B.G.; et al. ESCMID COVID-19 living guidelines: Drug treatment and clinical management. *Clin. Microbiol. Infect.* **2022**, *28*, 222–238. [CrossRef] [PubMed]
3. Wagner, C.; Griesel, M.; Mikolajewska, A.; Mueller, A.; Nothacker, M.; Kley, K.; Metzendorf, M.I.; Fischer, A.L.; Kopp, M.; Stegemann, M.; et al. Systemic corticosteroids for the treatment of COVID-19. *Cochrane Database Syst. Rev.* **2021**, *8*, CD014963. [PubMed]

4. Ansems, K.; Grundeis, F.; Dahms, K.; Mikolajevska, A.; Thieme, V.; Piechotta, V.; Metzendorf, M.I.; Stegemann, M.; Benstroem, C.; Fichtner, F. Remdesivir for the treatment of COVID-19. *Cochrane Database Syst. Rev.* **2021**, *8*, CD014962. [PubMed]
5. WHO. Solidarity Trial Consortium. Remdesivir and three other drugs for hospitalised patients with COVID-19: Final results of the WHO Solidarity randomised trial and updated meta-analyses. *Lancet* **2022**, *399*, 1941–1953. [CrossRef]
6. Kaka, A.S.; MacDonald, R.; Linskens, E.J.; Langsetmo, L.; Vela, K.; Duan-Porter, W.; Wilt, T.J. Major Update 2: Remdesivir for Adults With COVID-19: A Living Systematic Review and Meta-analysis for the American College of Physicians Practice Points. *Ann. Intern. Med.* **2022**, *175*, 701–709. [CrossRef]
7. Russo, A.; Binetti, E.; Borrazzo, C.; Gentilini Cacciola, E.; Battistini, L.; Ceccarelli, G.; Mastroianni, C.M.; d'Ettorre, G. Efficacy of Remdesivir-Containing Therapy in Hospitalized COVID-19 Patients: A Prospective Clinical Experience. *J. Clin. Med.* **2021**, *10*, 3784. [CrossRef]
8. Tejada, D.; Juanbeltz, R.; Rivero, M.; San Miguel, R.; Capdevila, F.; Beloqui, J.J.; Sarobe, M. Clinical course of patients with severe COVID-19 pneumonia treated with remdesivir: A real-life study. *PLoS ONE* **2022**, *17*, e0267283. [CrossRef]
9. Cogliati Dezza, F.; Oliva, A.; Mauro, V.; Romani, F.E.; Aronica, R.; Savelloni, G.; Casali, E.; Valeri, S.; Cancelli, F.; Mastroianni, C.M. Real-life use of remdesivir-containing regimens in COVID-19: A retrospective case-control study. *Infez. Med* **2022**, *30*, 211–222.
10. Goldberg, E.; Zvi, H.B.; Sheena, L.; Sofer, S.; Krause, I.; Sklan, E.H.; Shlomai, A. A real-life setting evaluation of the effect of remdesivir on viral load in COVID-19 patients admitted to a large tertiary centre in Israel. *Clin. Microbiol. Infect.* **2021**, *27*, 917.e1–917.e4. [CrossRef]
11. Soriano, A.; Montejano, R.; Sanz-Moreno, J.; Figueira, J.C.; Grau, S.; Güerri-Fernández, R.; Castro-Gómez, A.; Pérez-Román, I.; Hidalgo-Vega, A.; González-Domínguez, A. Impact of Remdesivir on the Treatment of COVID-19 During the First Wave in Spain. *Adv. Ther.* **2021**, *38*, 4057–4069. [CrossRef] [PubMed]
12. Garcia-Vidal, C.; Meira, F.; Cózar-Llistó, A.; Dueñas, G.; Puerta-Alcalde, P.; Garcia-Pouton, N.; Chumbita, M.; Cardozo, C.; Hernandez-Meneses, M.; Alonso-Navarro, R.; et al. COVID19-researcher group. Real-life use of remdesivir in hospitalized patients with COVID-19. *Rev. Esp. Quimioter.* **2022**, *34*, 136–140. [CrossRef] [PubMed]
13. Simioli, F.; Nicoletta, C.; Valentino, M.R.; Martino, M.; Annunziata, A.; Carannante, N.; Di Micco, P.; Fiorentino, G. Remdesivir in Severe COVID-19 and Non-Invasive Ventilation: A Real-Life Experience. *Healthcare* **2021**, *9*, 1108. [CrossRef] [PubMed]
14. Poliseno, M.; Gallo, C.; Cibelli, D.C.; Minafra, G.A.; Bottalico, I.F.; Bruno, S.R.; D'Errico, M.L.; Montemurro, L.; Rizzo, M.; Barbera, L.; et al. Efficacy and Safety of Remdesivir over Two Waves of the SARS-CoV-2 Pandemic. *Antibiotics* **2021**, *10*, 1477. [CrossRef] [PubMed]
15. Falcone, M.; Suardi, L.R.; Tiseo, G.; Barbieri, C.; Giusti, L.; Galfo, V.; Forniti, A.; Caroselli, C.; Della Sala, L.; Tempini, S.; et al. Early Use of Remdesivir and Risk of Disease Progression in Hospitalized Patients With Mild to Moderate COVID-19. *Clin. Ther.* **2022**, *44*, 364–373. [CrossRef]
16. Oliver, M.E.; Hinks, T.S.C. Azithromycin in viral infections. *Rev. Med. Virol.* **2021**, *31*, e2163. [CrossRef]
17. Ayerbe, L.; Risco-Risco, C.; Forgnone, I.; Pérez-Piñar, M.; Ayis, S. Azithromycin in patients with COVID-19: A systematic review and meta-analysis. *J. Antimicrob. Chemother.* **2022**, *77*, 303–309. [CrossRef]
18. Kamel, A.M.; Monem, M.S.A.; Sharaf, N.A.; Magdy, N.; Farid, S.F. Efficacy and safety of azithromycin in COVID-19 patients: A systematic review and meta-analysis of randomized clinical trials. *Rev. Med. Virol.* **2022**, *32*, e2258. [CrossRef]
19. Vaughn, V.M.; Gandhi, T.N.; Petty, L.A.; Patel, P.K.; Prescott, H.C.; Malani, A.N.; Ratz, D.; McLaughlin, E.; Chopra, V.; Flanders, S.A. Empiric Antibacterial Therapy and Community-onset Bacterial Coinfection in Patients Hospitalized with Coronavirus Disease 2019 (COVID-19): A Multi-hospital Cohort Study. *Clin. Infect. Dis.* **2021**, *72*, e533–e541. [CrossRef]
20. Lai, A.; Bergna, A.; Menzo, S.; Zehender, G.; Caucci, S.; Ghisetti, V.; Rizzo, F.; Maggi, F.; Cerutti, F.; Giurato, G.; et al. Collaborative Group SCIRE SARS-CoV-2 Italian Research Enterprise. Circulating SARS-CoV-2 variants in Italy, October 2020–March 2021. *Virol. J.* **2021**, *18*, 168. [CrossRef]
21. Meschi, T.; Rossi, S.; Volpi, A.; Ferrari, C.; Sverzellati, N.; Brianti, E.; Fabi, M.; Nouvenne, A.; Ticinesi, A. Reorganization of a large academic hospital to face COVID-19 outbreak: The model of Parma, Emilia-Romagna region, Italy. *Eur. J. Clin. Investig.* **2020**, *50*, e13250. [CrossRef]
22. Ticinesi, A.; Nouvenne, A.; Cerundolo, N.; Parise, A.; Prati, B.; Guerra, A.; Meschi, T. Trends of COVID-19 Admissions in an Italian Hub during the Pandemic Peak: Large Retrospective Study Focused on Older Subjects. *J. Clin. Med.* **2021**, *10*, 1115. [CrossRef] [PubMed]
23. Colombi, D.; Villani, G.D.; Maffi, G.; Risoli, C.; Bodini, F.C.; Petrini, M.; Morelli, N.; Anselmi, P.; Milanese, G ; Silva, M.; et al. Qualitative and quantitative chest CT parameters as predictors of specific mortality in COVID-19 patients. *Eur. Radiol.* **2020**, *27*, 701–710. [CrossRef] [PubMed]
24. World Health Organization. Living Guidance for Clinical Management of COVID-19. Version 23 November 2021. Available online: https://apps.who.int/iris/bitstream/handle/10665/349321/WHO-2019-nCoV-clinical-2021.2-eng.pdf (accessed on 7 July 2022).
25. De Terwangne, C.; Laouni, J.; Jouffe, L.; Lechien, J.R.; Bouillon, V.; Place, S.; Capulzini, L.; Machayekhi, S.; Ceccarelli, A.; Saussez, S.; et al. EPIBASE TEAM. Predictive Accuracy of COVID-19 World Health Organization (WHO) Severity Classification and Comparison with a Bayesian-Method-Based Severity Score (EPI-SCORE). *Pathogens* **2020**, *9*, 880. [CrossRef]
26. AIFA Agenzia Italiana del Farmaco. Trattamenti Utilizzabili nei Pazienti COVID-19 Nel Setting Ospedaliero. Versione 9/12/2020. Available online: www.aifa.gov.it (accessed on 16 June 2022). (In Italian)

27. AIFA Agenzia Italiana del Farmaco. Remdesivir Nella Terapia dei Pazienti Adulti con COVID-19. Versione 24/11/2020. Available online: www.aifa.gov.it (accessed on 16 June 2022). (In Italian)
28. Beigel, J.H.; Tomashek, K.M.; Dodd, L.E.; Mehta, A.K.; Zingman, B.S.; Kalil, A.C.; Hohmann, E.; Chu, H.Y.; Luetkemeyer, A.; Kline, S.; et al. ACTT-1 Study Group Members. Remdesivir for the Treatment of COVID-19—Final Report. *N. Engl. J. Med.* **2020**, *383*, 1813–1826. [CrossRef] [PubMed]
29. Paranjape, N.; Husain, M.; Priestley, J.; Koonjah, Y.; Watts, C.; Havlik, J. Early Use of Remdesivir in Patients Hospitalized with COVID-19 Improves Clinical Outcomes: A Retrospective Observational Study. *Infect. Dis. Clin. Pract.* **2021**, *29*, e282–e286. [CrossRef] [PubMed]
30. Alsayed, A.A.H.; Sharif-Askari, F.S.; Sharif-Askari, N.S.; Hussain, A.A.S.; Hamid, Q.; Halwani, R. Early administration of remdesivir to COVID-19 patients associates with higher recovery rate and lower need for ICU admission: A retrospective cohort study. *PLoS ONE* **2021**, *16*, e0258643.
31. Gottlieb, R.L.; Vaca, C.E.; Paredes, R.; Mera, J.; Webb, B.J.; Perez, G.; Oguchi, G.; Ryan, P.; Nielsen, B.U.; Brown, M.; et al. GS-US-540-9012 (PINETREE) Investigators. Early Remdesivir to Prevent Progression to Severe COVID-19 in Outpatients. *N. Engl. J. Med.* **2022**, *386*, 305–315. [CrossRef]
32. Du, X.; Zuo, X.; Meng, F.; Han, C.; Ouyang, W.; Han, Y.; Gu, Y.; Zhao, X.; Xu, F.; Qin, F.X. Direct inhibitory effect on viral entry of influenza A and SARS-CoV-2 viruses by azithromycin. *Cell. Prolif.* **2021**, *54*, e12953. [CrossRef]
33. Fiolet, T.; Guihur, A.; Rebeaud, M.E.; Mulot, M.; Peiffer-Smadja, M.; Mahamat-Saleh, Y. Effect of hydroxychloroquine with or without azithromycin on the mortality of coronavirus disease 2019 (COVID-19) patients: A systematic review and meta-analysis. *Clin. Microbiol. Infect.* **2021**, *27*, 19–27. [CrossRef]
34. Prediletto, I.; D'Antoni, L.; Carbonara, P.; Daniele, F.; Dongilli, R.; Flore, R.; Pacilli, A.M.G.; Pisani, L.; Tomsa, C.; Vega, M.L.; et al. Standardizing PaO_2 for $PaCO_2$ in P/F ratio predicts in-hospital mortality in acute respiratory failure due to COVID-19: A pilot prospective study. *Eur. J. Intern. Med.* **2021**, *92*, 48–54. [CrossRef] [PubMed]
35. Geriatric Medicine Research Collaborative; COVID Collaborative; Welch, C. Age and frailty are independently associated with increased COVID-19 mortality and increased care needs in survivors: Results of an international multi-centre study. *Age Ageing* **2021**, *50*, 617–630. [PubMed]
36. Dumitrascu, F.; Branje, K.E.; Hladkowicz, E.S.; Lalu, M.; McIsaac, D.I. Association of frailty with outcomes in individuals with COVID-19: A living review and meta-analysis. *J. Am. Geriatr. Soc.* **2021**, *69*, 2419–2429. [CrossRef] [PubMed]
37. Ticinesi, A.; Nouvenne, A.; Prati, B.; Guida, L.; Parise, A.; Cerundolo, N.; Bonaguri, C.; Aloe, R.; Guerra, A.; Meschi, T. The Clinical Significance of Procalcitonin Elevation in Patients over 75 Years Old Admitted for COVID-19 Pneumonia. *Mediat. Inflamm.* **2021**, *2021*, 5593806. [CrossRef]
38. Ticinesi, A.; Nouvenne, A.; Parise, A.; Prati, B.; Meschi, T. Defining SARS-CoV-2 breakthrough infection needing hospitalization in mass vaccination era: From disease-centered to patient-centered care. *Acta Biomed.* **2022**, *93*, e2022182.

antibiotics

MDPI

Article

Safety and Efficacy of Ivermectin for the Prevention and Treatment of COVID-19: A Double-Blinded Randomized Placebo-Controlled Study

Nasikarn Angkasekwinai [1,*], Pinyo Rattanaumpawan [1], Methee Chayakulkeeree [1], Pakpoom Phoompoung [1], Pornpan Koomanachai [1], Sorawit Chantarasut [2], Walaiporn Wangchinda [1], Varalak Srinonprasert [3,4] and Visanu Thamlikitkul [1]

1 Division of Infectious Diseases and Tropical Medicine, Department of Medicine, Faculty of Medicine Siriraj Hospital, Mahidol University, 2 Wanglang Road, Bangkoknoi, Bangkok 10700, Thailand; pinyo.rat@mahidol.ac.th (P.R.); methee.cha@mahidol.ac.th (M.C.); pakpoom.pho@mahidol.ac.th (P.P.); pornpan.koo@mahidol.ac.th (P.K.); walaiporn.wan@mahidol.ac.th (W.W.); visanu.tha@mahidol.ac.th (V.T.)
2 Department of Medicine, Faculty of Medicine Siriraj Hospital, Mahidol University, 2 Wanglang Road, Bangkoknoi, Bangkok 10700, Thailand; sorawit.cha@mahidol.ac.th
3 Division of Geriatric Medicine, Department of Medicine, Faculty of Medicine Siriraj Hospital, Mahidol University, 2 Wanglang Road, Bangkoknoi, Bangkok 10700, Thailand; varalak sri@mahidol.ac.th
4 Siriraj Research Data Management Unit (Si-RDMU), Department of Research, Faculty of Medicine Siriraj Hospital, Mahidol University, 2 Wanglang Road, Bangkoknoi, Bangkok 10700, Thailand
* Correspondence: nasikarn@gmail.com or nasikarn.ang@mahidol.ac.th; Tel.: +66-2-419-9462; Fax: +66-419-7783

Abstract: The safety and efficacy of ivermectin for the prevention and treatment of COVID-19 are still controversial topics. From August to November 2021, we conducted a double-blinded, randomized controlled trial at Siriraj Hospital, Thailand. Eligible participants were adults $\geq$ 18 years with suspected COVID-19 who underwent a SARS-CoV-2 RT-PCR test. After enrollment, the participants were randomized to receive either ivermectin (400–600 μg/kg/d) or placebo once daily for 3 days. Among 983 participants, 536 (54.5%) with a negative RT-PCR result were enrolled in the prevention study, and 447 (45.5%) with a positive RT-PCR result were enrolled in the treatment study. In the prevention study, the incidence of COVID-19 on Day 14 was similar between the ivermectin and the placebo group (4.7% vs. 5.2%; $p = 0.844$; $\Delta = -0.4\%$; 95% CI; −4.3–3.5%). In the treatment study, there was no significant difference between the ivermectin and placebo group for any Day 14 treatment outcome: proportion with oxygen desaturation (2.7% vs. 1.9%; $p = 0.75$), change in WHO score from baseline (1 [−5, 1] vs. 1 [−5, 1]; $p = 0.50$), and symptom resolution (76% vs. 82.2%; $p = 0.13$). The ivermectin group had a significantly higher proportion of transient blurred vision (5.6% vs. 0.6%; $p < 0.001$). Our study failed to demonstrate the efficacy of a 3-day once daily of ivermectin for the prevention and treatment of COVID-19. The given regimen of ivermectin should not be used for either prevention or treatment of COVID-19 in populations with a high rate of COVID-19 vaccination.

Keywords: COVID-19; ivermectin; randomized-controlled trial; prevention; treatment

Citation: Angkasekwinai, N.; Rattanaumpawan, P.; Chayakulkeeree, M.; Phoompoung, P.; Koomanachai, P.; Chantarasut, S.; Wangchinda, W.; Srinonprasert, V.; Thamlikitkul, V. Safety and Efficacy of Ivermectin for the Prevention and Treatment of COVID-19: A Double-Blinded Randomized Placebo-Controlled Study. *Antibiotics* **2022**, *11*, 796. https://doi.org/10.3390/antibiotics11060796

Academic Editors: Sabine Danielle Allard and Johan Van Laethem

Received: 24 May 2022
Accepted: 11 June 2022
Published: 12 June 2022

1. Background

The coronavirus disease 2019 (COVID-19) pandemic, caused by severe acute respiratory syndrome coronavirus 2 (SARS-CoV-2), remains a major health threat, with almost 440 million confirmed cases and six million deaths globally as of 1 March 2022 [1]. Effective vaccines are an essential measure to limit the COVID-19 pandemic; however, breakthrough infections and the continuation of the pandemic might occur owing to the emergence of new SARS-CoV-2 variants of concern [2]. The availability of effective antiviral treatments remains limited. Repurposing existing medicines that are readily available and inexpensive is therefore of great interest [3].

Ivermectin, an oral antiparasitic agent with broad-spectrum antiviral activity, has shown potent in vitro anti-SARS-CoV-2 activity; it induced a 5000-fold reduction in viral RNA after 48 h, with a half-maximal inhibitory concentration (IC_{50}) of 2 μM [4]. Because of its several mechanisms for potential antiviral and anti-inflammatory activity [5], ivermectin has been evaluated in many studies for the treatment of SARS-CoV-2 infection, with high doses of up to 2 mg/kg and courses of up to 4 days [6,7]. However, most previous studies on the efficacy of ivermectin to treat COVID-19 were non-randomized and open-label, performed in settings with limited access to COVID-19 vaccines [8–10]. Furthermore, there were only few studies that focused on the efficacy of ivermectin for prevention of SARS-CoV-2 infection [11]. To determine the efficacy of ivermectin in preventing the acquisition of SARS-CoV-2 among a high-risk exposure population, and to evaluate the efficacy of ivermectin for treating laboratory-confirmed COVID-19, we performed a pragmatic randomized, placebo-controlled trial comparing a 3-day once daily dose of ivermectin with a placebo in an outpatient setting.

2. Methods

2.1. *Study Design and Patients*

This was a single-center, double-blinded, pragmatic randomized placebo-controlled trial conducted from August to November 2021 at Siriraj Hospital, a 2300-bed university hospital in Bangkok, Thailand. The study protocol was approved by the Siriraj Institutional Review Board (certificate of approval no. Si 607/2021) and was conducted in accordance with the Declaration of Helsinki. Written informed consent was obtained from all participants.

Participants were eligible for inclusion if they were ≥18 years and were suspected of having SARS-CoV-2 infection because of their respiratory tract symptoms or because they had a history of contact with a confirmed COVID-19 patient. Eligible participants also must have had a documented positive or negative test for SARS-CoV-2 (RT-PCR) from a nasopharyngeal (NP) swab sample taken on the enrollment day. Participants were excluded if they were pregnant or breastfeeding, had a history of ivermectin hypersensitivity, had a previous SARS-CoV-2 infection within 3 months, or had an inconclusive result on their RT-PCR SARS-CoV-2 test.

2.2. *Randomization and Masking*

Participants were randomized in a 1:1 ratio to receive standard of care plus ivermectin (Atlantic Laboratory Ltd., Bangkok, Thailand) or an identical placebo. Randomization was performed by using the computer-generated method with a varying block size of 2 to 8. Only the pharmacist knew the treatment assignment. The participants and investigators were blinded to the treatment assignment for the entire study period.

2.3. *Interventions*

Participants were given either placebo or ivermectin based on their body weight; the ivermectin dose ranged from 400–600 μg/kg/d. The dosage was calculated to the nearest 6 mg or 12 mg whole tablets (dosing table in the study protocol, Supplement File S1). The participants were advised to take the study medication before a meal on the enrollment day (Day 0) and once every 24 h for 2 consecutive days. After the RT-PCR result was available (within the same day), the participants with a negative result were included into the prevention study, while those with a positive result were included into the treatment study.

2.4. *Procedures*

Participants in the prevention study were instructed to collect an NP swab for the rapid detection of SARS-CoV-2 antigen using the Standard Q COVID-19 Ag test (SD Biosensor, Inc., Gyeonggi-do, Yongin-si, Korea) on Day 14 and whenever they developed new symptoms suggestive of COVID-19. If the rapid antigen test was positive, NP swab sampling for RT-PCR testing was performed at the hospital. Participants in the treatment study were instructed to measure their temperature and oxygen saturation on Day 3, Day 7, and Day 14. In accordance with the Thailand National Clinical Practice Guidelines for the Treatment of COVID-19, favipiravir was recommended for all symptomatic patients and all asymptomatic patients with risk factors for disease progression.

All participants were contacted by telephone on Day 3, Day 7, and Day 14 to collect data on temperature, oxygen saturation, symptoms, and safety of study medication.

2.5. Outcome Measurement

The primary outcomes of both prevention and treatment studies were analyzed using intention to treat (ITT) and modified intention to treat (mITT) populations. The ITT population comprised all eligible participants who were randomized and applied a worse-case scenario. All participant without evaluable outcomes and drop-out participant were considered as having a poor outcome. The mITT population included all randomized participants who received at least one dose of study drug. Participants in the prevention study who did not perform a second NP swab within 14 days were assumed to have a negative RT-PCR result in the mITT population if they were asymptomatic on Day 28 without proof of a RT-PCR test taken elsewhere.

The primary outcome of the prevention study was the proportion of participants with a positive RT-PCR within 14 days after enrollment among those with a negative RT-PCR result at enrollment in the mITT population. The primary outcomes of the treatment study were the proportion of participants with oxygen desaturation (oxygen saturation < 96% or decreased from baseline by $\geq$3% after exertion); changes in the WHO 10-point clinical progression score [12] on Day 3, Day 7, and Day 14 compared to baseline; the absence of all symptoms at Day 3, Day 7, and Day 14; hospitalization within 14 days; and 28-day mortality in the mITT population.

The secondary outcome of the study was the safety of the study medications, including the number and the percentage of participants with adverse effects (AEs) evaluated in the mITT population.

2.6. Sample Size

For the primary outcomes of the prevention study, we anticipated that a 3-day course of ivermectin would reduce the rate of SARS-CoV-2 infection from 20% to 10%. To achieve a power of 80% and a two-sided p-value of 0.05, 199 participants/group were required. Considering potential dropouts, a total of 478 participants with a negative RT-PCR at the enrollment were needed.

For the primary outcome of the treatment study, it was assumed that ivermectin would reduce the rate of oxygen desaturation of COVID-19 patients from 30% to 15%. To achieve 80% power and a two-sided p-value of 0.05, 121 participants/group were required. Considering potential dropouts, 290 participants with a positive RT-PCR at the enrollment were needed.

Given that the prevalence of SARS-CoV-2 infection among the patients who visited the acute respiratory tract infection (ARI) clinic was 50%, we needed to enroll at least 1000 patients who presented to the ARI clinic to achieve the target sample size for both studies.

2.7. Statistical Analysis

Demographic and baseline characteristics are presented as descriptive statistics. Continuous data are presented as the mean (standard deviation) or median (range), as appropriate. Categorical data are presented as number and percentage. The unpaired t-test and Mann–Whitney U test were used to compare continuous data, while the chi-square test or Fisher's exact test was used to compare categorical data as appropriate. All statistical analyses were performed with PASW Statistics (SPSS) 18.0 (IBM Corp., Armonk, NY, USA). A p-value < 0.05 was considered statistically significant.

3. Results

Among 1236 patients who were screened from August 2021–October 2021, 1000 were recruited for the study; 500 were randomized to receive ivermectin and 500 to receive placebo. Seventeen participants (1.7%) were excluded owing to the pre-specified exclusion criteria. Among 983 participants, 968 (98.5%) completed the 28-day follow-up (Figure 1). The baseline information and clinical characteristics of the 983 participants were similar between two groups (Table S1). The mean age of all participants was 38.4 ± 12.1 years, 57.4% were female, and 30.6% had pre-existing diseases. Overall, 80% of the participants had previously received ≥1 dose of a COVID-19 vaccine. Of the 983 participants, 536 (54.5%) with a negative RT-PCR SARS-CoV-2 test were included in the prevention study, and 447 (45.5%) with a positive RT-PCR were included in the treatment study (Figure 1).

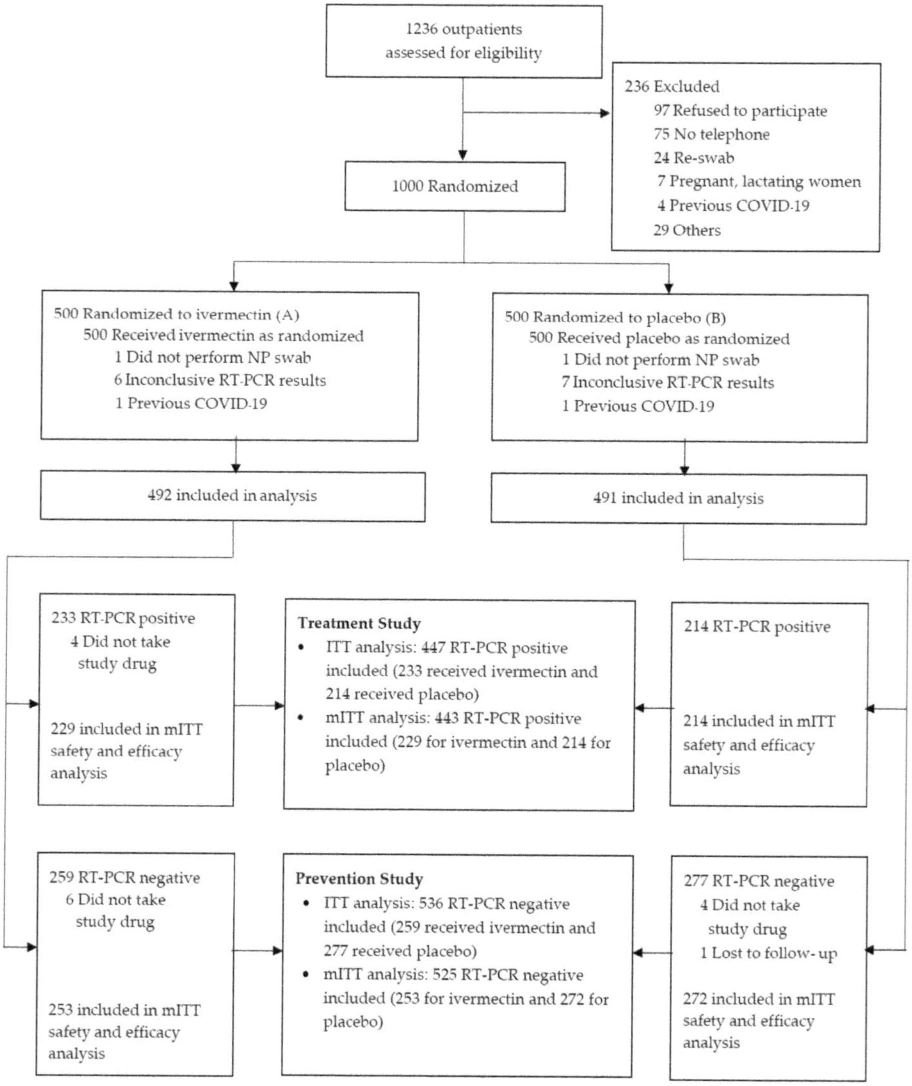

Figure 1. Enrollment, randomization, and treatment assignment.

3.1. Primary Outcome of Ivermectin Prevention Study

Among the 536 participants with a negative RT-PCR result at enrollment, 259 were in the ivermectin group and 277 were in the placebo group. The baseline and clinical characteristics of the participants in both groups were similar (Table 1) The mean age was 37.6 ± 12.0 years, 57.8% were female, and 29.5% had pre-existing diseases. Approximately 90% of the participants had exposure risk, mainly a household contact with confirmed COVID-19, within 7 days before their RT-PCR test. Nearly 40% of participants were asymptomatic, and most (85%) had previously received ≥1 dose of a COVID-19 vaccine. Of the 536 participants, 11 participants were excluded from the mITT analysis because of various reasons. Therefore, 525 participants (253 in the ivermectin group and 272 in the placebo group) were included in the mITT analysis (Figure 1). Three participants in the ivermectin group and two participants in the placebo group did not perform follow-up NP swab testing for SARS-CoV-2 detection. The proportion of positive RT-PCR within 14 days was similar in the ivermectin and placebo groups for the ITT analysis (6.95% vs. 6.86%, p = 1.000), with a difference of −0.09% [95%CI, −4.3–4.6%]. The proportions were also similar in the mITT analysis (4.74% vs. 5.15%, p = 0.844), with a difference of −0.41% [95%CI, −4.3–3.5%] (Table 2). The median time to a positive RT-PCR test was 6 days, and there was no significant difference between the groups. In the mITT population subgroup analyses, there were no differences in the proportion of participants with a positive RT-PCR when analyzed by the contact duration, body weight, and vaccination status (Table S2).

Table 1. Baseline characteristics of participants with a negative RT-PCR at enrollment in the ivermectin prevention study.

Characteristics	Total	Ivermectin	Placebo	p Value
	n = 536	n = 259	n = 277	
Age (years)				
Mean (SD)	37.6 (12.0)	37.8 (12.6)	37.4 (11.6)	0.727
Median (range)	37 (18, 72)	37 (18, 72)	37 (18, 60)	0.960
Gender, n (%)				0.930
Male	226 (42.2)	110 (42.2)	116 (42.0)	
Female	310 (57.8)	149 (57.5)	161 (58.1)	
Body weight, kg				
Median (range)	65.1 (35.3, 142.5)	64.4 (35.3, 142.5)	65.3 (37.8, 110.2)	0.995
Mean (SD)	67.0 (15.9)	67.3 (16.9)	66.7 (14.9)	0.672
≤90 kg	487 (90.9)	232 (89.6)	255 (92.1)	0.369
>90 kg	49 (9.1)	27 (10.4)	22 (7.9)	
Presence of underlying diseases, n (%)	158 (29.5)	73 (28.2)	85 (30.7)	0.570
Hypertension	47 (8.8)	20 (7.7)	27 (9.7)	0.447
Diabetes mellitus	25 (4.7)	10 (3.9)	15 (5.4)	0.420
Dyslipidemia	25 (4.7)	10 (3.9)	15 (5.4)	0.420
Coronary artery disease	6 (1.1)	5 (1.9)	1 (0.4)	0.112
Chronic lung diseases	1 (0.2)	0 (0.0)	1 (0.4)	1.000
Cerebrovascular disease	2 (0.4)	0 (0.0)	2 (0.7)	0.500
Cancer	7 (1.3)	3 (1.2)	4 (1.4)	1.000
Others	97 (18.1)	45 (17.4)	52 (18.8)	0.737
Duration between last exposure to a COVID-19 patient and enrollment (n = 495)				1.000
Median (range)	2 (0, 66)	2 (0, 17)	3 (0, 66)	0.336
≤7 days	443 (89.5)	219 (89.4)	224 (89.6)	1.000
>7 days	52 (10.5)	26 (10.6)	26 (10.4)	
Exposure risk: household contact	495 (92.4)	245 (94.6)	250 (90.3)	0.073
Presence of symptoms, n (%)				
Asymptomatic	206 (38.4)	104 (40.2)	102 (36.8)	0.477
Symptomatic	330 (61.6)	155 (59.8)	175 (63.2)	

Table 1. *Cont.*

Characteristics	Total	Ivermectin	Placebo	*p* Value
	n = 536	*n* = 259	*n* = 277	
Sore throat	186 (34.7)	91 (35.1)	95 (34.3)	0.856
Cough	136 (25.4)	57 (22.0)	79 (28.5)	0.092
Runny nose	89 (16.6)	43 (16.6)	46 (16.6)	1.000
Fever	73 (13.6)	32 (12.4)	41 (14.8)	0.451
Dyspnea	27 (5.0)	9 (3.5)	18 (6.5)	0.118
Diarrhea	19 (3.5)	9 (3.5)	10 (3.6)	1.000
Chest pain	6 (1.1)	2 (0.8)	4 (1.4)	0.687
Vomiting	5 (0.9)	4 (1.5)	1 (0.4)	0.202
Loss of taste/smell	4 (0.7)	0 (0.0)	4 (1.4)	0.124
Others	114 (21.3)	52 (20.1)	62 (22.4)	0.528
Duration of illness, (*n* = 330)				
Median (range)	2 (0, 20)	2 (0, 20)	2 (0, 14)	0.692
<3 days	191 (57.9)	88 (56.8)	103 (58.9)	0.738
≥3 days	139 (42.1)	67 (43.2)	72 (41.1)	
Previous COVID-19 vaccination, *n* (%)				0.604
No	85 (15.9)	45 (17.4)	40 (14.4)	
Incomplete vaccine course (1 dose with last dose < 2 weeks prior)	34 (6.3)	13 (5.0)	21 (7.6)	
Incomplete vaccine course (1 dose with last dose ≥ 2 weeks prior)	185 (34.5)	92 (35.5)	93 (33.6)	
Completed vaccine course (2 doses with last dose < 2 weeks prior)	64 (11.9)	32 (12.4)	32 (11.6)	
Completed vaccine course (2 doses with last dose ≥ 2 weeks prior or 3 doses with any duration)	168 (31.3)	77 (29.7)	91 (32.9)	
Compliance with study medication				0.884
Full compliance, *n* (%)	485 (90.5)	235 (90.7)	250 (90.3)	
Partial compliance, *n* (%)	51 (9.5)	24 (9.3)	27 (9.7)	

SD: standard deviation.

Table 2. Primary outcomes of the ivermectin prevention study classified by ITT and mITT analyses.

Primary Outcomes	Ivermectin	Placebo	*p* Value
ITT analysis (*n* = 536)	*n* = 259	*n* = 277	
Proportion of COVID-19 infection within 14 days, *n* (%)	18 (6.95)	19 (6.86)	1.000
Difference (95% CI)	0.09% (−4.30–4.57)		
Median (range) time to positive SARS-CoV-2 test (days)	6 (3, 11)	6 (1, 14)	0.327
Modified ITT analysis (*n* = 525)	*n* = 253	*n* = 272	
Proportion of COVID-19 infection within 14 days, *n* (%)	12 (4.74)	14 (5.15)	0.844
Difference (95% CI)	−0.41% (−4.28–3.53)		
Median (range) time to positive SARS-CoV-2 test (days)	6 (3, 11)	4.5 (1, 14)	0.374
Ct value of participants who became positive within 14 days, mean (SD) *			
N gene	18.0 (2.8)	16.8 (3.0)	0.418
E gene	14.3 (2.9)	13.3 (3.0)	0.456
RdRp gene	18.9 (2.8)	18.1 (2.8)	0.674

* The Ct data were available for only 22 participants (10 in ivermectin group and 12 in placebo group). Four participants who became RT-PCR positive were tested at another hospital.

3.2. Primary Outcomes of Ivermectin Treatment Study

Among the 447 participants with a positive RT-PCR at enrollment, 233 were in the ivermectin group and 214 were in the placebo group. The baseline and clinical characteristics were similar in the groups (Table 3). The mean age was 39.5 ± 12.1 years, 56.8% were female, and 32% had pre-existing diseases. Approximately 88% of participants had ≥1 symptom, of which cough (50.6%), sore throat (47%), and fever (38%) were the most frequent. Overall,

55.6% of participants had onset of symptoms ≤ 3 days before enrollment, and 60% of participants had a cycle threshold (Ct) value ≤ 20. Overall, 21.5% of the participants were COVID-19 vaccine-naive. Almost all (97.5%) received favipiravir concomitantly with the study medication. Four participants in the ivermectin group were excluded because they did not take the drug. Therefore, 443 participants (229 in the ivermectin group and 214 in the placebo group) were included in the mITT analysis (Figure 1).

Table 3. Baseline characteristics of participants with a positive RT-PCR at enrollment in the ivermectin treatment study.

Characteristics	Total	Ivermectin	Placebo	*p* Value
	n = 447	*n* = 233	*n* = 214	
Age (years)				
Mean (SD)	39.5 (12.1)	39.1 (12.0)	39.8 (12.3)	0.570
Median (range)	39 (18, 72)	39 (18, 69)	40 (18, 72)	0.612
Gender, *n* (%)				0.566
Male	193 (43.2)	104 (44.6)	89 (41.6)	
Female	254 (56.8)	129 (55.4)	125 (58.4)	
Body weight, kg				
Median (range)	66.2 (36.3, 138.0)	66.3 (36.3, 138.0)	66.2 (36.6, 118.5)	0.598
Mean (SD)	68.5 (16.1)	68.1 (16.3)	69.0 (15.9)	0.608
≤90 kg, *n* (%)	406 (90.8)	214 (91.8)	192 (89.7)	0.512
>90 kg, *n* (%)	41 (9.2)	19 (8.2)	22 (10.3)	
Presence of underlying diseases, *n* (%)	143 (32.0)	70 (30.0)	73 (34.1)	0.363
Hypertension	50 (11.2)	22 (9.4)	28 (13.1)	0.233
Diabetes mellitus	31 (6.9)	14 (6.0)	17 (7.9)	0.460
Dyslipidemia	25 (5.6)	12 (5.2)	13 (6.1)	0.686
Coronary artery disease	8 (1.8)	4 (1.7)	4 (1.9)	1.000
Chronic kidney disease	2 (0.4)	1 (0.4)	1 (0.5)	1.000
Cirrhosis	1 (0.2)	1 (0.4)	0 (0.0)	1.000
Chronic lung diseases	1 (0.2)	0 (0.0)	1 (0.5)	0.481
Cerebrovascular disease	1 (0.2)	0 (0.0)	1 (0.5)	0.481
Cancer	1 (0.2)	0 (0.0)	1 (0.5)	0.481
Autoimmune disease	2 (0.4)	0 (0.0)	2 (0.9)	0.229
Others	62 (13.9)	36 (15.5)	26 (12.1)	0.340
Exposure risk: household contact, *n* (%)	314 (70.2)	158 (67.8)	156 (72.9)	0.256
Duration between last exposure to a COVID-19 patient and enrollment (*n* = 313)				
Median (range)	2 (0, 25)	2.5 (0, 25)	2 (0, 16)	0.356
≤7 days, *n* (%)	292 (93.3)	146 (92.4)	146 (94.2)	0.653
>7 days, *n* (%)	21 (6.7)	12 (7.6)	9 (5.8)	
Presence of symptoms, *n* (%)				
Asymptomatic	52 (11.6)	24 (10.3)	28 (13.1)	0.379
Symptomatic	395 (88.4)	209 (89.7)	186 (86.9)	
Cough	226 (50.6)	129 (55.4)	97 (45.3)	0.037
Sore throat	210 (47.0)	115 (49.4)	95 (44.4)	0.299
Fever	170 (38.0)	90 (38.6)	80 (37.4)	0.845
Runny nose	156 (34.9)	85 (36.5)	71 (33.2)	0.488
Loss of taste/smell	79 (17.7)	34 (14.6)	45 (21.0)	0.083
Dyspnea	31 (6.9)	21 (9.0)	10 (4.7)	0.093
Diarrhea	25 (5.6)	12 (5.2)	13 (6.1)	0.686
Chest pain	5 (1.1)	2 (0.9)	3 (1.4)	0.674
Vomiting	2 (0.4)	1 (0.4)	1 (0.5)	1.000
Others	126 (28.2)	70 (30.0)	56 (26.2)	0.400
Duration of illness, (*n* = 394)				
Median (range)	2 (0, 10)	2 (0, 10)	2 (0, 10)	0.990
<3 days, *n* (%)	219 (55.6)	115 (55.3)	104 (55.9)	0.919
≥3 days, *n* (%)	175 (44.4)	93 (44.7)	82 (44.1)	
RT-PCR Ct value				

Table 3. *Cont.*

Characteristics	Total	Ivermectin	Placebo	*p* Value
	n = 447	*n* = 233	*n* = 214	
Mean (SD)	20.2 (5.3)	20.0 (5.2)	20.4 (5.4)	0.460
<20, *n* (%)	266 (59.5)	141 (60.5)	125 (58.4)	0.700
≥20, *n* (%)	181 (40.5)	92 (39.5)	89 (41.6)	
Oxygen saturation (%), mean (SD)	97.9 (1.1)	97.9 (1.0)	97.9 (1.2)	0.964
Oxygen saturation < 96%, *n* (%)	6 (1.3)	1 (0.4)	5 (2.3)	0.109
WHO clinical score, median (range)	2 (1, 2)	2 (1, 2)	2 (1, 2)	0.360
Score 1, *n* (%)	52 (11.6)	24 (10.3)	28 (13.1)	0.379
Score 2, *n* (%)	395 (88.4)	209 (89.7)	186 (86.9)	
Previous vaccination, *n* (%)				0.522
No	112 (25.1)	65 (27.9)	47 (22.0)	
Incomplete vaccine course (1 dose with last dose < 2 weeks prior)	30 (6.7)	17 (7.3)	13 (6.1)	
Incomplete vaccine coursee (1 dose with last dose ≥ 2 weeks prior)	184 (41.2)	90 (38.6)	94 (43.9)	
Completed vaccine coursee (2 doses with last dose < 2 weeks prior)	25 (5.6)	11 (4.7)	14 (6.5)	
Completed vaccine coursee (2 doses with last dose ≥ 2 weeks prior or 3 doses with any duration)	96 (21.5)	50 (21.5)	46 (21.5)	
Chest X-ray, *n* (%)				0.993
Normal	264 (59.1)	138 (59.2)	126 (58.9)	
Unilateral infiltrate	7 (1.6)	4 (1.7)	3 (1.4)	
Bilateral infiltrate	6 (1.3)	3 (1.3)	3 (1.4)	
Not done	170 (38.0)	88 (37.8)	82 (38.3)	
Admission type at baseline, *n* (%)				0.072
Quarantine hotel	280 (62.6)	145 (62.2)	135 (63.1)	
Home isolation	132 (29.5)	67 (28.8)	65 (30.4)	
Hospital	33 (7.4)	21 (9.0)	12 (5.6)	
No admission	1 (0.2)	0 (0.0)	1 (0.5)	
Unknown	1 (0.2)	0 (0.0)	1 (0.5)	
Concomitant medication, *n* (%)				
Favipiravir	435 (97.5)	226 (97.4)	209 (97.7)	1.000
Others	3 (0.7)	0 (0.0)	3 (1.4)	0.110
Compliance with study medication, *n* (%)				0.762
Full compliance	399 (89.3)	209 (89.7)	190 (88.8)	
Partial compliance	48 (10.7)	24 (10.3)	24 (11.2)	

Ct: cycle threshold; RT-PCR: reverse transcription-polymerase chain reaction; SD: standard deviation; WHO: World Health Organization.

For both the ITT and mITT analyses, there were no significant differences between ivermectin (plus favipiravir) and the placebo (plus favipiravir) for all outcomes, including the proportion of participants with oxygen desaturation; the change in WHO progression score from baseline; the absence of symptoms at Day 3, Day 7, and Day 14; 14-day hospitalization rate; and 28-day mortality (Table 4). Most symptoms gradually subsided over time except for loss of smell, which showed a peak frequency on Day 3 (Figure S1). In the mITT population, subgroup analysis did not reveal any differences in outcomes between the ivermectin group and the placebo group (Table S3). No participants died in this study. One participant in the ivermectin group and one participant in the placebo group reported COVID-19 infection on Day 23 and Day 17, respectively. In addition, no factors associated with favorable outcomes in participants who had an absence of all symptoms on Day 7 could be identified (Table S4).

Table 4. Primary outcomes of the ivermectin treatment study classified by ITT and mITT analyses.

Primary Outcomes	Ivermectin	Placebo	*p* Value
ITT analysis (n = 447)	n = 233	n = 214	
Proportion of participants with oxygen desaturation, n (%) **			
Day 3	2 (0.9)	3 (1.4)	0.674
Day 7	2 (0.9)	4 (1.9)	0.433
Day 14	6 (2.6)	4 (1.9)	0.753
Change in WHO progression score from baseline			
Day 3	0 (−3, 0)	0 (−5, 0)	0.462
Day 7	0 (−4, 0)	0 (−5, 0)	0.256
Day 14	1 (−5, 1)	1 (−5, 1)	0.348
Absence of all symptoms, n (%)			
Day 3	57 (24.5)	44 (20.6)	0.365
Day 7	118 (50.6)	115 (53.7)	0.570
Day 14	174 (74.7)	176 (82.2)	0.066
Hospitalization due to clinical progression within 14 days, n (%)	8 (3.4)	4 (1.9)	0.386
28-day mortality	0	0	-
Modified ITT analysis (n = 443)	n = 229	n = 214	
Proportion of participants with oxygen desaturation, n (%) **			
Day 3	2 (0.9)	3 (1.4)	0.676
Day 7	2 (0.9)	4 (1.9)	0.435
Day 14	6 (2.7)	4 (1.9)	0.752
Change in WHO progression score from baseline			
Day 3	0 (−3, 0)	0 (−5, 0)	0.436
Day 7	0 (−4, 0)	0 (−5, 0)	0.239
Day 14	1 (−5, 1)	1 (−5, 1)	0.501
Absence of all symptoms, n (%)			
Day 3	56 (24.5)	44 (20.6)	0.364
Day 7	118 (51.5)	115 (53.7)	0.703
Day 14	174 (76.0)	176 (82.2)	0.129
Hospitalization due to clinical progression within 14 days, n (%)	4 (1.7)	4 (1.9)	1.000
28-day mortality	0	0	-

** Oxygen desaturation refers to oxygen saturation < 96% or a decrease in oxygen saturation ≥ 3% after exertion; CI: confidence interval; Ct: cycle threshold; ITT: intention to treat; SD: standard deviation; WHO: World Health Organization.

3.3. *Adverse Events*

The incidences of AEs in participants in both groups are shown in Table 5. There was no significant difference in the proportion of participants reporting AEs between the ivermectin and placebo groups (21.6% vs. 18.9%, p = 0.337). However, there were more ocular AEs reported in the ivermectin group (5.6% vs. 0.6%, p < 0.001). These were mainly blurred vision while taking ivermectin, but this spontaneously resolved after completing the medication. An analysis of AEs by the study cohort found that ocular problems were more prevalent in the ivermectin group than the placebo group in the treatment cohort (8.7% vs. 0%, p < 0.001). Headache was reported more often in the placebo group (4.5% vs. 1.9%, p = 0.027) (Table S5). No serious AEs were reported in this study.

Table 5. Adverse events reported by all participants in the ivermectin prevention and treatment studies.

AEs (mITT Population)	Ivermectin (n = 482)		Placebo (n = 486)		p Value
	No. Events	No. Cases n (%)	No. Events	No. Cases n (%)	
Total	141	104 (21.6)	144	92 (18.9)	0.337
Ocular problems	28	27 (5.6)	4	3 (0.6)	<0.001
Diarrhea	23	23 (4.8)	21	19 (3.9)	0.532
Myalgia	15	13 (2.7)	19	17 (3.5)	0.579
Headache	10	9 (1.9)	25	22 (4.5)	0.027
Neurologic symptoms	8	8 (1.7)	11	10 (2.1)	0.813
Rash	7	7 (1.5)	4	4 (0.8)	0.383
Nausea/vomiting	6	6 (1.2)	12	11 (2.3)	0.328
Pruritus	1	1 (0.2)	3	3 (0.6)	0.624
Others	43	40 (8.3)	45	44 (9.1)	0.732

AE: adverse event; mITT: modified intention to treat.

4. Discussion

To the best of our knowledge, this was the first large, double-blinded, randomized controlled trial to determine the safety and efficacy of ivermectin for both the treatment and prevention of COVID-19 in the same outpatient setting. A high dose of ivermectin (400–600 μg/kg/d) for 3 days did not show a significant benefit for the prevention of SARS-CoV-2 infection. Similarly, early treatment with the same dose and duration of ivermectin did not reduce disease progression or hospitalization in patients with mild-to-moderate COVID-19 compared with the placebo group. No serious AEs were reported in this study. However, eye-related symptoms, particularly blurred vision, occurred more frequently in the ivermectin group, especially in those who concomitantly received favipiravir.

In this study, there was a low rate of acquiring SARS-CoV-2 infection (5%) even though the study was conducted among people with high-risk exposure. Ivermectin did not show a benefit for preventing SARS-CoV-2 infection, which is in contrast with some previous studies. A recent open-labeled randomized study evaluated 303 asymptomatic household contacts in Egypt found that the proportion of clinically diagnosed SARS-CoV-2 infections was 7.4% in the ivermectin group and 58.5% in the control group [13]. Another matched case–control study conducted in India among 186 healthcare workers who received two doses of 300 μg/kg ivermectin 3 days apart found a 73% reduction in SARS-CoV-2 infection in the following month [11]. However, these previous studies were non-randomized studies with subjective outcome measurement. The low rate of a positive RT-PCR within 14 days in our study could have several explanations. First, our study was conducted after several months of a national COVID-19 vaccination campaign; therefore, 85% of participants had already received ≥1 dose of COVID-19 vaccine. Second, all confirmed COVID-19 cases in Thailand were requested to self-quarantine at home or in designated facilities to prevent further transmission [14]. This might have resulted in the low COVID-19 incidence rates in the study.

Our study demonstrated that early treatment with ivermectin did not reduce COVID-19 disease progression or the hospitalization rate and did not increase symptom resolution.

Several randomized controlled trials on the efficacy of ivermectin for treating COVID-19 have shown conflicting results in terms of virological and clinical outcomes [15,16]. However, our study results were in line with several well-controlled studies. The study conducted in Colombia did not find any clinical benefit of a 10-day ivermectin therapy among mild-to-moderate COVID-19 cases [17]. The IVERCOR-COVID19 study did not find any benefit of ivermectin therapy on preventing hospitalization [18], and a recent study in Brazil evaluating the efficacy of 3-day ivermectin for mild-to-moderate COVID-19 with risk factors also did not reduce the rate of hospitalization within 28 days compared with placebo (14.7% vs. 16.3%, respectively) [19]. The results of studies investigating ivermectin as a COVID-19 treatment may depend on the study quality [20].

The lack of observed differences in clinical outcomes between ivermectin and placebo in our treatment study should not be related with using favipiravir as standard of care. From recent systematic reviews, favipiravir did not show a significant benefit on the viral clearance and mortality [21]. In addition, it is possible that our study population had a low rate of outcomes because only one-third of the participants had a co-morbidity, which may have resulted in a lower rate of disease progression.

An adequate and safe dose of ivermectin for treating COVID-19 has not been clearly established. Ivermectin's IC_{50} against SARS-CoV-2 was found to be 2 μM, which is >35 times higher than the maximal plasma concentration after oral ivermectin administration at the approved dose of 200 μg/kg [4]. The present study used a higher daily dose (400–600 μg/kg/d) than the standard regimen, aiming to achieve a high drug concentration during peak viremia; this dosage was found to be safe and well tolerated in a previous study [22]. However, the previous pharmacokinetic (PK) study showed that an ivermectin dosage of 10 times higher than the approved dose was not sufficient to reach the required IC50 in the lungs [23]. A recent study using a high dose of ivermectin (600 μg/kg/d) for 5 days did not reduce the SARS-CoV-2 viral load [24]. Our study did not show any benefit in clinical endpoints from high-dose ivermectin (400–600 μg/kg/d for 3 days), which is in line with these previous PK and clinical studies.

The significantly higher rate of transient blurred vision in the ivermectin group has been documented. The previous malaria study reported a significant high rate of transient visual disturbance: 8% among those who receive a moderate dose of ivermectin (300 μg/kg for 3 days) and 22% among those who received a high dose of ivermectin (600 μg/kg ivermectin for 3 days) [22]. The transient visual disturbance was possibly due to ivermectin potentiating GABA release and binding, resulting in central nervous system AEs such as mydriasis. Importantly, ocular adverse events were significantly more prevalent if co-administration with favipiravir. Nevertheless, further investigation is required to confirm the possibility of ivermectin–favipiravir drug interaction and ocular AEs. It is unclear why there was a lower rate of headache in the ivermectin group. This might have occurred by chance.

Our study has some limitations. In the ivermectin prevention study, we used self-conducted rapid antigen testing to determine the presence of SARS-CoV-2, and only those with a positive rapid antigen test underwent confirmation testing by RT-PCR. However, participants using NP swab sampling for their self-conducted test could have obtained a false-negative rapid antigen result because of improper collection technique or test performance. However, a distribution of this phenomenon should have occurred similarly in both groups. In addition, the incidence of COVID-19 infection in the prevention study was much lower than we expected. This might be due to several factors, such as the majority of participants in our study have received at least one dose of COVID-19 vaccination, or the changes in the SARSCoV-2 strain from time to time. To detect the difference of the small effect in the prevention study between ivermectin and placebo, more than 3200 participants may be required. Therefore, the result of this prevention study warrants further, larger research.

5. Conclusions

In this double-blinded, pragmatic randomized placebo-controlled trial, ivermectin did not demonstrate a protective effect for preventing SARS-CoV-2 infection. The results also showed that ivermectin had no COVID-19 therapeutic effect in combination with standard of care (favipiravir). Transient blurred vision was significantly more common in participants who received ivermectin plus favipiravir. Therefore, ivermectin should not be used for preventing SARS-CoV-2 infection or for treating mild-to-moderate COVID-19.

Supplementary Materials: The following supporting information can be downloaded at: https://www.mdpi.com/article/10.3390/antibiotics11060796/s1, Supplementary File S1: Study protocol Ivermectin COVID; Figure S1: Details of symptoms by day of follow-up among participants with COVID-19 in the ivermectin treatment study; Table S1: Baseline characteristics of all study participants; Table S2: Primary outcomes in subgroup population of ivermectin prevention study by analysis of mITT population; Table S3: Primary outcomes in subgroup population of ivermectin treatment study by analysis of mITT population; Table S4: Factors associated with favorable outcome at Day 7 in treatment study; Table S5: Adverse events of study medication separated by prevention or treatment study. References [4,12,18,25–31] are cited in the supplementary materials.

Author Contributions: Conceptualization, N.A., P.R., M.C., P.P., P.K., W.W. and V.T.; Data Curation, N.A. and S.C.; Methodology; N.A., P.R., M.C., P.P., P.K., W.W., V.S. and V.T.; Validation, N.A., S.C., V.S. and V.T.; Formal Analysis, N.A., P.R., M.C., V.S. and V.T.; Investigation, N.A., P.R., M.C., P.P., S.C., V.S. and V.T.; Resources, N.A., V.S. and V.T.; Writing—Review and Editing, N.A., P.R., M.C., P.P., P.K. and V.T.; Visualization, N.A., P.R., M.C., P.P. and V.T.; Supervision, N.A., V.S., V.T. and N.A.; Project Administration, N.A., S.C., V.S. and V.T.; Funding Acquisition, N.A. All authors commented on previous versions of the manuscript. All authors have read and agreed to the published version of the manuscript.

Funding: This study was supported by Siriraj Foundation, Faculty of Medicine Siriraj Hospital, Mahidol University, Bangkok, Thailand (grant no. D004039). Atlantic Laboratory Ltd., Bangkok, Thailand provided the ivermectin and placebo used in this study. The funding source had no role in the study design; in data collection, analysis, or interpretation; in the conclusions drawn; or in the preparation of the manuscript.

Institutional Review Board Statement: The study was conducted according to the guidelines of the Declaration of Helsinki and was approved by the Siriraj Institutional Review Board (certificate of approval no. Si 607/2021).

Informed Consent Statement: Written informed consent was obtained from all participants.

Data Availability Statement: The datasets generated and/or analyzed during the current study are available from the corresponding author on reasonable request.

Acknowledgments: The authors gratefully acknowledge all study team members: the healthcare personnel at the acute respiratory tract infection clinic for their cooperation, the research nurses for data collection, and Julaporn Pooliam for statistical analysis. The authors also gratefully thank all participants for participating in the study, as well as Katherine Thieltges for editing a draft of this manuscript.

Conflicts of Interest: The authors declare no conflict of interest.

References

1. World Health Organization. *WHO Coronavirus (COVID-19) Dashboard*; World Health Organization: Geneva, Switzerland, 2022. Available online: https://covid19.who.int (accessed on 1 March 2022).
2. Murray, C.J.L.; Piot, P. The Potential Future of the COVID-19 Pandemic: Will SARS-CoV-2 Become a Recurrent Seasonal Infection? *JAMA* **2021**, *325*, 1249–1250. [CrossRef] [PubMed]
3. Rayner, C.R.; Dron, L.; Park, J.J.H.; Decloedt, E.H.; Cotton, M.F.; Niranjan, V.; Smith, P.F.; Dodds, M.G.; Brown, F.; Reis, G.; et al. Accelerating Clinical Evaluation of Repurposed Combination Therapies for COVID-19. *Am. J. Trop. Med. Hyg.* **2020**, *103*, 1364–1366. [CrossRef] [PubMed]
4. Caly, L.; Druce, J.D.; Catton, M.G.; Jans, D.A.; Wagstaff, K.M. The FDA-approved drug ivermectin inhibits the replication of SARS-CoV-2 in vitro. *Antivir. Res.* **2020**, *178*, 104787. [CrossRef] [PubMed]
5. Zhang, X.; Song, Y.; Ci, X.; An, N.; Ju, Y.; Li, H.; Wang, X.; Han, C.; Cui, J.; Deng, X. Ivermectin inhibits LPS-induced production of inflammatory cytokines and improves LPS-induced survival in mice. *Inflamm. Res.* **2008**, *57*, 524–529. [CrossRef] [PubMed]
6. Popp, M.; Stegemann, M.; Metzendorf, M.I.; Gould, S.; Kranke, P.; Meybohm, P.; Skoetz, N.; Weibel, S. Ivermectin for preventing and treating COVID-19. *Cochrane Database Syst. Rev.* **2021**, *7*, CD015017. [CrossRef]
7. Navarro, M.; Camprubi, D.; Requena-Mendez, A.; Buonfrate, D.; Giorli, G.; Kamgno, J.; Gardon, J.; Boussinesq, M.; Munoz, J.; Krolewiecki, A. Safety of high-dose ivermectin: A systematic review and meta-analysis. *J. Antimicrob. Chemother.* **2020**, *75*, 827–834. [CrossRef]

8. Abd-Elsalam, S.; Noor, R.A.; Badawi, R.; Khalaf, M.; Esmail, E.S.; Soliman, S.; Abd El Ghafar, M.S.; Elbahnasawy, M.; Moustafa, E.F.; Hassany, S.M.; et al. Clinical study evaluating the efficacy of ivermectin in COVID-19 treatment: A randomized controlled study. *J. Med. Virol.* **2021**, *93*, 5833–5838. [CrossRef]
9. Ahmed, S.; Karim, M.M.; Ross, A.G.; Hossain, M.S.; Clemens, J.D.; Sumiya, M.K.; Phru, C.S.; Rahman, M.; Zaman, K.; Somani, J.; et al. A five-day course of ivermectin for the treatment of COVID-19 may reduce the duration of illness. *Int. J. Infect. Dis.* **2021**, *103*, 214–216. [CrossRef]
10. Ravikirti; Roy, R.; Pattadar, C.; Raj, R.; Agarwal, N.; Biswas, B.; Manjhi, P.K.; Rai, D.K.; Shyama; Kumar, A.; et al. Evaluation of Ivermectin as a Potential Treatment for Mild to Moderate COVID-19: A Double-Blind Randomized Placebo Controlled Trial in Eastern India. *J. Pharm. Pharm. Sci.* **2021**, *24*, 343–350. [CrossRef]
11. Behera, P.; Patro, B.K.; Singh, A.K.; Chandanshive, P.D.; Ravikumar, S.R.; Pradhan, S.K.; Pentapati, S.S.K.; Batmanabane, G.; Mohapatra, P.R.; Padhy, B.M.; et al. Role of ivermectin in the prevention of SARS-CoV-2 infection among healthcare workers in India: A matched case-control study. *PLoS ONE* **2021**, *16*, e0247163. [CrossRef]
12. WHO Working Group on the Clinical Characterisation and Management of COVID-19 Infection. A minimal common outcome measure set for COVID-19 clinical research. *Lancet Infect. Dis.* **2020**, *20*, e192–e197. [CrossRef]
13. Shoumann, W.M.; Hegazy, A.A.; Nafae, R.M.; Ragab, M.I.; Samra, S.R.; Ibrahim, D.A.; Al-Mahrouky, T.H.; Sileem, A.E. Use of Ivermectin as a Potential Chemoprophylaxis for COVID-19 in Egypt: A Randomised Clinical Trial. *J. Clin. Diagn. Res.* **2021**, *15*, 6. [CrossRef]
14. World Health Organization. Joint Intra-Action Review of the Public Health Response to COVID-19 in Thailand, 20–24 July 2020. Available online: https://www.who.int/docs/default-source/searo/thailand/iar-covid19-en.pdf (accessed on 1 March 2022).
15. Bartoletti, M.; Azap, O.; Barac, A.; Bussini, L.; Ergonul, O.; Krause, R.; Pano-Pardo, J.R.; Power, N.R.; Sibani, M.; Szabo, B.G.; et al. ESCMID COVID-19 living guidelines: Drug treatment and clinical management. *Clin. Microbiol. Infect.* **2022**, *28*, 222–238. [CrossRef] [PubMed]
16. Buonfrate, D.; Chesini, F.; Martini, D.; Roncaglioni, M.C.; Ojeda Fernandez, M.L.; Alvisi, M.F.; De Simone, I.; Rulli, E.; Nobili, A.; Casalini, G.; et al. High-dose ivermectin for early treatment of COVID-19 (COVER study): A randomised, double-blind, multicentre, phase II, dose-finding, proof-of-concept clinical trial. *Int. J. Antimicrob. Agents* **2022**, *56*, 106516. [CrossRef] [PubMed]
17. Lopez-Medina, E.; Lopez, P.; Hurtado, I.C.; Davalos, D.M.; Ramirez, O.; Martinez, E.; Diazgranados, J.A.; Onate, J.M.; Chavarriaga, H.; Herrera, S.; et al. Effect of Ivermectin on Time to Resolution of Symptoms Among Adults with Mild COVID-19: A Randomized Clinical Trial. *JAMA* **2021**, *325*, 1426–1435. [CrossRef]
18. Vallejos, J.; Zoni, R.; Bangher, M.; Villamandos, S.; Bobadilla, A.; Plano, F.; Campias, C.; Chaparro Campias, E.; Medina, M.F.; Achinelli, F.; et al. Ivermectin to prevent hospitalizations in patients with COVID-19 (IVERCOR-COVID19) a randomized, double-blind, placebo-controlled trial. *BMC Infect. Dis.* **2021**, *21*, 635. [CrossRef]
19. Reis, G.; Silva, E.; Silva, D.C.M.; Thabane, L.; Milagres, A.C.; Ferreira, T.S.; Dos Santos, C.V.Q.; Campos, V.H.S.; Nogueira, A.M.R.; de Almeida, A.; et al. Effect of Early Treatment with Ivermectin among Patients with Covid-19. *N. Engl. J. Med.* **2022**, *386*, 1721–1731. [CrossRef]
20. Hill, A.; Mirchandani, M.; Pilkington, V. Ivermectin for COVID-19: Addressing Potential Bias and Medical Fraud. *Open Forum. Infect. Dis.* **2022**, *9*, ofab645. [CrossRef]
21. Hassanipour, S.; Arab-Zozani, M.; Amani, B.; Heidarzad, F.; Fathalipour, M.; Martinez-de-Hoyo, R. The efficacy and safety of Favipiravir in treatment of COVID-19: A systematic review and meta-analysis of clinical trials. *Sci. Rep.* **2021**, *11*, 11022; Erratum in *Sci. Rep.* **2022**, *12*, 1996. [CrossRef]
22. Smit, M.R.; Ochomo, E.O.; Aljayyoussi, G.; Kwambai, T.K.; Abong'o, B.O.; Chen, T.; Bousema, T.; Slater, H.C.; Waterhouse, D.; Bayoh, N.M.; et al. Safety and mosquitocidal efficacy of high-dose ivermectin when co-administered with dihydroartemisinin-piperaquine in Kenyan adults with uncomplicated malaria (IVERMAL): A randomised, double-blind, placebo-controlled trial. *Lancet Infect. Dis.* **2018**, *18*, 615–626. [CrossRef]
23. Schmith, V.D.; Zhou, J.J.; Lohmer, L.R.L. The Approved Dose of Ivermectin Alone is not the Ideal Dose for the Treatment of COVID-19. *Clin. Pharmacol. Ther.* **2020**, *108*, 762–765. [CrossRef] [PubMed]
24. Krolewiecki, A.; Lifschitz, A.; Moragas, M.; Travacio, M.; Valentini, R.; Alonso, D.F.; Solari, R.; Tinelli, M.A.; Cimino, R.O.; Alvarez, L.; et al. Antiviral effect of high-dose ivermectin in adults with COVID-19: A proof-of-concept randomized trial. *EClinicalMedicine* **2021**, *37*, 100959. [CrossRef] [PubMed]
25. Available online: https://ddc.moph.go.th/viralpneumonia/index.php (accessed on 1 March 2022).
26. Sirijatuphat, R.; Suputtamongkol, Y.; Angkasekwinai, N.; Horthongkham, N.; Chayakulkeeree, M.; Rattanaumpawan, P.; Kantakamalakul, W. Epidemiology, clinical characteristics, and treatment outcomes of patients with COVID-19 at Thailand's university-based referral hospital. *BMC Infect. Dis.* **2021**, *21*, 382. [CrossRef] [PubMed]
27. Guzzo, C.A.; Furtek, C.I.; Porras, A.G.; Chen, C.; Tipping, R.; Clineschmidt, C.M.; Sciberras, D.G.; Hsieh, J.Y.K.; Lasseter, K.C. Safety, tolerability, and pharmacokinetics of escalating high doses of ivermectin in healthy adult subjects. *J. Clin. Pharmacol.* **2002**, *42*, 1122–1133. [CrossRef] [PubMed]
28. Cavalcanti, A.B.; Berwanger, O.; Zampieri, F.G. Hydroxychloroquine with or without Azithromycin in Covid-19. Reply. *N. Engl. J. Med.* **2021**, *384*, 191.
29. Roman, Y.M.; Burela, P.A.; Pasupuleti, V.; Piscoya, A.; Vidal, J.E.; Hernandez, A.V. Ivermectin for the treatment of COVID-19: A systematic review and meta-analysis of randomized controlled trials. *Clin. Infect. Dis.* **2021**, *28*, e434–e460.

30. Andrew Hill, A.A.; Ahmed, S.; Asghar, A. *Meta-Analysis of Randomized Trials of Ivermectin to Treat SARS-CoV-2 Infection*; Research Square: Durham, NC, USA, 2021.
31. Hellwig, M.D.; Maia, A.A. COVID-19 prophylaxis? Lower incidence associated with prophylactic administration of ivermectin. *Int. J. Antimicrob. Agents* **2021**, *57*, 106248. [CrossRef]

antibiotics

Article

A Retrospective, Monocentric Study Comparing Co and Secondary Infections in Critically Ill COVID-19 and Influenza Patients

Diane Marcoux [1,†], Isabelle Etienne [2,†], Alain Van Muylem [2], Elisa Gouvea Bogossian [3], Nicolas Yin [4], Fabio Silvio Taccone [3] and Maya Hites [1,*]

1 Clinic of Infectious Diseases, HUB-Erasme Hospital, 1070 Brussels, Belgium; diane.marcoux@chu-charleroi.be
2 Department of Pneumology, HUB-Erasme Hospital, 1070 Brussels, Belgium; isabelle.etienne@erasme.ulb.ac.be (I.E.); alain.van.muylem@erasme.ulb.ac.be (A.V.M.)
3 Department of Intensive Care, HUB-Erasme Hospital, 1070 Brussels, Belgium; elisagobog@gmial.com (E.G.B.); fabio.taccone@erasme.ulb.ac.be (F.S.T.)
4 Department of Microbiology, Laboratoire Hospitalier Universitaire de Bruxelles-Universitair Laboratorium Brussel (LHUB-ULB), Université Libre de Bruxelles, 1070 Brussels, Belgium; nicolas.yin@erasme.ulb.ac.be
* Correspondence: maya.hites@erasme.ulb.ac.be
† These authors contributed equally to this work.

Abstract: Few data are available on infectious complications in critically ill patients with different viral infections. We performed a retrospective monocentric study including all of the patients admitted to the intensive care unit (ICU) with confirmed COVID-19 (as of 13 March 2020) or Influenza A and/or B infections (as of 1 January 2015) until 20 April 2020. Coinfection and secondary infections (occurring within and after 48 h from admission, respectively) were recorded. Fifty-seven COVID-19 and 55 Influenza patients were included. Co-infections were documented in 13/57 (23%) COVID-19 patients vs. 40/55 (73%) Influenza patients ($p < 0.001$), most of them being respiratory (9/13, 69% vs. 35/40, 88%; $p = 0.13$) and of bacterial origin (12/13, 92% vs. 29/40, 73%; $p = 0.25$). Invasive aspergillosis infections were observed only in Influenza patients (8/55, 15%). The COVID-19 and Influenza patients presented 1 (0–4) vs. 0 (0–4) secondary infections ($p = 0.022$), with comparable sites being affected (lungs: 35/61, 57% vs. 13/31, 42%; $p = 0.16$) and causative pathogens occurring (Gram-negative bacteria: 51/61, 84% vs. 23/31, 74%; $p > 0.99$). The COVID-19 patients had longer ICU lengths of stay (15 (–65) vs. 5 (1–89) days; $p = 0.001$), yet the two groups had comparable mortality rates (20/57, 35% vs. 23/55, 41%; $p = 0.46$). We report fewer co-infections but more secondary infections in the ICU COVID-19 patients compared to the Influenza patients. Most of the infectious complications were respiratory and of bacterial origin.

Keywords: COVID-19; Influenza; co-infection; secondary infections; aspergillosis

Citation: Marcoux, D.; Etienne, I.; Van Muylem, A.; Bogossian, E.G.; Yin, N.; Taccone, F.S.; Hites, M. A Retrospective, Monocentric Study Comparing Co and Secondary Infections in Critically Ill COVID-19 and Influenza Patients. *Antibiotics* **2022**, *11*, 704. https://doi.org/10.3390/antibiotics11060704

Academic Editors: Sabine Danielle Allard and Johan Van Laethem

Received: 22 April 2022
Accepted: 20 May 2022
Published: 24 May 2022

1. Introduction

In December 2019, a new virus named Severe Acute Respiratory Syndrome Coronavirus 2 (SARS-CoV-2) was identified in Wuhan city, Hubei Province, China, causing a new disease called COVID-19, which can lead to severe acute respiratory distress syndrome (ARDS) with a relatively high risk of death [1]. This virus quickly spread worldwide, resulting in a global health crisis and a rapid saturation of health care services, including intensive care units (ICU) [2].

Several studies have already been published concerning the occurrence of co-infections and/or secondary infections in ICU COVID-19 patients [3–11]. Most of these studies are retrospective and report data on small cohorts including heterogenous populations. There are only two published meta-analyses based on retrospective studies: one that included small cohorts from six studies only focusing on ICU patients and eight studies focusing on both ICU and non-ICU patients that reported on COVID-19 co-infections [11]; and one that included five studies focusing only on ICU patients that described co-infections (four of those studies) and secondary infections (one of those studies) [3]. According to these studies, the rate of co-infections varied from 4.3% to 28% [3–9], and the rate of secondary infections ranged from 40.7% to 51% [4,7–10]. It is unclear whether the risk of coinfections or secondary infections is similar to that of other viral infections.

Influenza affects up to 20% of the population each year. Bacterial and fungal co-infections are frequent occurrences among ICU patients with Influenza infection. Co-infected patients have higher morbidity and mortality rates than those without co-infections [12]. The mechanisms of co-pathogenesis have been studied and are multifactorial [12,13]. First, the virus causes epithelial dysfunction. The ciliary damage prevents efficient pathogen clearance, and the epithelial cells and surfactant destruction give the pathogens access to nutrients. Second, these structural damages, together with the local inflammation, facilitate bacterial adherence to their specific receptors. Bacteria also produce pro-inflammatory cytokines resulting in the synergistic activation of immune response. Third, the immune response to the influenza virus interferes with the normal pathogen recognition effector responses during and after the viral infection, enabling secondary bacterial infections [14].

A large French retrospective study comparing 89,530 COVID-19 patients to 45,819 Influenza patients reported a higher in-hospital death rate among the COVID-19 patients requiring ICU admission compared to the Influenza patients; however, the rates of coinfections and secondary infections were not reported [15]. Another retrospective study comparing 642 COVID-19 patients to 742 Influenza patients reported more secondary bacterial infections in the COVID-19 patients, which was an independent predictor of death in the COVID-19 patients [16]. In a retrospective study, Bardi and al. also reported a higher ICU mortality in secondary infected COVID-19 ICU patients compared to those not infected [9].

This study therefore aims to describe whether the occurrence, type, and outcome of co-infections and secondary infections in ICU patients are different between the COVID-19 and Influenza diseases.

2. Results

2.1. Patients' Characteristics

A total of 57 COVID-19 patients and 55 Influenza A or B patients were eligible for the analysis over the study period. The characteristics of the study population are presented in Table 1.

Table 1. Baseline patient characteristics.

	COVID-19	**Influenza**	***p*-Value**
	N = 57	**N = 55**	
Sex—male	41 (72)	30 (54)	0.056
Age (years)	61 (53–70)	65 (54–77)	0.108
Comorbidities			
Current smokers, n (%)	6 (10)	23 (42)	<0.001
Obesity	n = 54; 19 (35)	n = 46; 14 (29)	0.534
BMI (kg/m^2)	n = 54; 28 (24–31)	n = 46; 28 (24–33)	0.841
Arterial hypertension, n (%)	37 (65)	34 (62)	0.734
Chronic cardiomyopathy, n (%)	9 (16)	23 (42)	0.002
Chronic pulmonary disease, n (%)	17 (30)	27 (49)	0.037
Chronic kidney failure, n (%)	4 (7)	13 (24)	0.014
Diabetes, n (%)	11 (19)	16 (29)	0.226
Active neoplasia, n (%)	3 (5)	3 (5)	0.999
Charlson Comorbidity Index, n (%)	3 (1–5)	4 (3–6)	0.077
Immunosuppressive therapy, n (%)	6 (10)	24 (44)	<0.001
Chronic steroids, n (%)	4 (7)	16 (29)	0.002
Solid organ transplant, n (%)	3 (5)	6 (11)	0.317

Continuous variables are reported as the median (interquartile ranges), and categorical variables are reported as counts (percentages). A $p < 0.05$ was considered statistically significant. Abbreviation—BMI: body mass index.

The patients were predominantly men in both groups and had comparable ages (63 years (53–74)), Charlson Comorbidity Indexes (4 (2–5)), and prevalence of obesity (32%). The COVID-19 patients were less frequently smokers and had less chronic cardiomyopathy, pulmonary disease, and kidney failure than the Influenza patients. They were also significantly less immunocompromised than the Influenza patients, receiving chronic steroid treatment less frequently.

2.2. Clinical Data upon ICU Admission

The clinical data upon ICU admission are presented in Table 2a. The patients were mainly admitted to the ICU for respiratory failure. The patients were severely ill, with comparable baseline SOFA and SAPS3 scores and PaO_2/FiO_2 ratios in both patient groups (medians of 6 (3–10); 58 (48–68) and 145 (106–213), respectively). The delay between the hospital and ICU admission was significantly shorter in the COVID-19 group compared to that in the Influenza group. Upon admission, the COVID-19 patients presented significantly more ground glass features on the CT-scan and a lower leucocyte count, but they presented a higher lymphocyte count and higher c-reactive protein and lactate dehydrogenase values than the Influenza patients. Significantly fewer COVID-19 patients received antibiotics before and during the first 48 h of ICU admission compared to the Influenza cohort.

Table 2. (a). Clinical data upon ICU admission. (b). Microbiological data upon ICU admission.

(a)

	COVID-19	Influenza	*p*-Value
	N = 57	N = 55	
Reason for ICU admission			
Respiratory failure, n (%)	51 (89)	47 (86)	0.52
Medical reason, n (%)	6 (11)	7 (13)	0.716
Surgical reason, n (%)	0 (0)	1 (2)	0.477
Hospital to ICU admission, days	1 (0–4)	2 (0–11)	0.012
Baseline SOFA	7 (3–9)	6 (3–10)	0.943
Baseline SAPS 3	55 (46–68)	62 (52–71)	0.07
Baseline PaO_2/FiO_2 ratio	140 (100–194)	152 (112–213)	0.432
CT features	n = 48	n = 27	
Ground glass, n (%)	46 (96)	15 (56)	<0.001
Condensations, n (%)	28 (58)	19 (73)	0.209
Biological data			
Leukocytes, $10^3/mm^3$	8.69 (6.52–11.09)	11.03 (8.42–15.00)	0.035
Lymphocytes count, $10^3/mm^3$	0.89 (0.57–1.32)	0.54 (0.34–1.04)	0.028
CRP, mg/L	144 (94–230)	63 (41–125)	<0.001
Creatinine, mg/dL	1.1 (0.8–1.6)	1.2 (0.9–1.8)	0.522
LDH, IU/L	506 (360–620)	303 (239–431)	<0.001
Antibiotherapy before ICU admission, n (%)	18 (32)	30 (54)	0.04
Antibiotherapy during the first 48 h, n (%)	22 (39)	44 (80)	<0.001

(b)

	COVID-19	Influenza	*p*-Value
	N = 57	N = 55	
Bacteriological samples within 48 h of admission			
Total respiratory samples, n (%)	34 (60)	46 (84)	0.005
Sputum or ETA, n (%)	29 (51)	36 (67)	0.091
BAL, n (%)	25 (44)	12 (22)	0.016
Multiplex respiratory PCR panel, n (%)	23 (40)	21 (39)	0.875
Influenza test (PCR or Ag), n (%)	42 (74)	55 (100)	<0.001
Blood cultures, n (%)	54 (95)	36 (66)	<0.001
Urine cultures, n (%)	54 (95)	20 (38)	<0.001

Table 2. *Cont.*

(b)			
	COVID-19	**Influenza**	***p*-Value**
	N = 57	**N = 55**	
Bacteriological data			
Total co-infections, n (%)	13 (23)	40 (73)	<0.001
Respiratory co-infections, n (%) *	9 (16)	35 (64)	<0.001
Bacteremia, n (%) *	2 (3)	7 (13)	0.091
Urinary tract infection, n (%) *	2 (3)	1 (2)	0.317
Pathogens of documented co-infections **			
Gram-positive coccus, n (%)	7 (54)	18 (45)	0.579
Staphylococcus aureus, n (%)	1 (14)	9 (50)	
Streptococcus pneumoniae, n (%)	3 (43)	5 (28)	
Other *Streptococcus* spp., n (%)	3 (43)	3 (17)	
Other, n (%)	0 (0)	1 (5)	
Gram-negative bacillus, n (%)	9 (69)	16 (40)	0.607
Escherichia coli, n (%)	3 (33)	1 (6)	
Klebsiella spp., n (%)	1 (11)	4 (25)	
Pseudomonas aeruginosa, n (%)	0 (0)	1 (6)	
Haemophilus influenzae, n (%)	3 (33)	5 (31)	
Other, n (%)	2 (22)	5 (31)	
Virus, n (%)	2 (14)	14 (35)	0.181
Adenovirus, n (%)	2 (100)	0 (0)	
Cytomegalovirus, n (%)	0 (0)	4 (29)	
Coronavirus (other than COVID-19), n (%)	0 (0)	4 (29)	
Other, n (%)	0 (0)	8 (57)	
Aspergillus sp., n (%)	0 (0)	8 (20)	0.002

Continuous variables are reported as the median (interquartile range), and categorical variables are reported as numbers (percentages). A $p < 0.05$ was considered statistically significant. Abbreviations—ICU: Intensive Care Unit, SOFA: Sequential Organ Failure Assessment, SAPS 3: Simplified Acute Physiology Score 3, PaO_2/FiO_2 ratio: ratio of partial oxygen pressure to the fraction inspired air, CRP: C-reactive protein, LDH: lactate dehydrogenase, ETA: endotracheal aspirations, BAL: bronchoalveolar lavage. Categorical variables are reported as numbers (percentages). A $p < 0.05$ was considered statistically significant. Abbreviations—ETA: endotracheal aspiration, PCR: polymerase chain reaction. * Co-infections could be multisite. ** Co-infections were sometimes polymicrobial.

2.3. Co-Infections

The data on co-infections are reported in Table 2b and Figure 1. The microbiological documentation was more systematic in the COVID-19 patients than in the Influenza patients upon ICU admission, meaning that significantly more BAL, blood, and urine cultures were performed for the COVID-19 patients compared to the Influenza patients. Nevertheless, the COVID-19 patients presented significantly fewer co-infections than the Influenza patients. The co-infections were predominantly respiratory in both groups. The most frequent pathogens were Gram-positive cocci (*Streptococcus pneumoniae* in the COVID-19 group and *Staphylococcus aureus* in the Influenza group) and Gram-negative rods (*Escherichia coli* in the COVID-19 group and *Haemophilus influenza* in the Influenza group). Only two COVID-19 patients were diagnosed with a viral co-infection (14%), whereas 14 (35%) were diagnosed in the Influenza group ($p = 0.181$). No co-infections were observed with *Aspergillus* sp. in the COVID-19 group, whereas 8 (20%) were observed in the Influenza group, ($p = 0.002$). Seven of those patients had chronic pulmonary disease, five were receiving chronic steroid treatment, and one was a solid organ transplant recipient. The median time to ICU admission for these patients with invasive aspergillosis infections was 15 days (12–20), which is significantly longer than the median time to ICU admission for the entire cohort: 1.5 days (0–6).

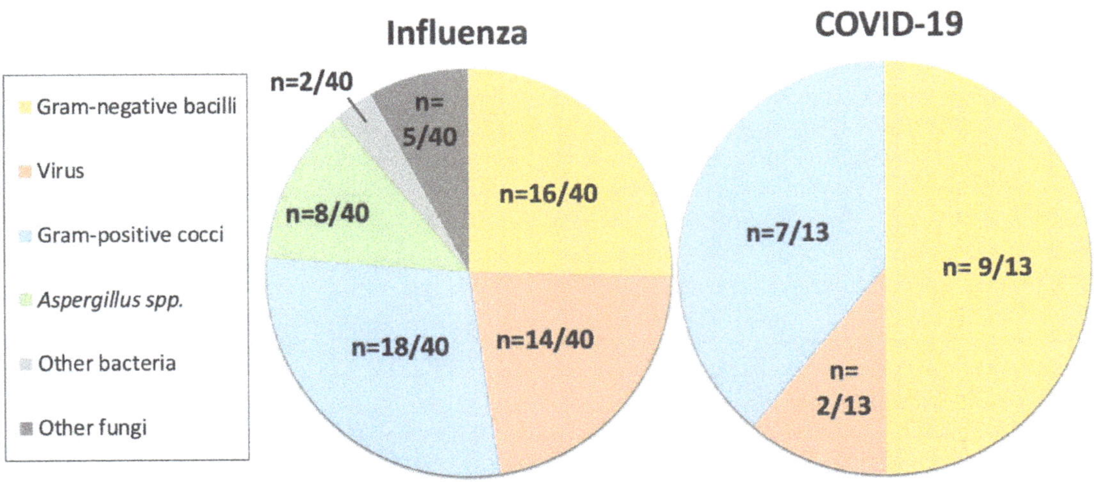

Figure 1. Co-infections in COVID-19 patients versus Influenza ICU patients.

Table 3 provides the results of the univariate logistic regression model exploring the risk factors for co-infections. Table S1 provides the frequencies of each variable among the co-infected and non-co-infected patients in both cohorts. Influenza infection and immunosuppression (IS) therapy were identified as risk factors for co-infections upon ICU admission. Each variable for which the univariate odds ratio (OR) yielded a p-value ≤ 0.1 was then included in a multivariable logistic regression model. Only IS therapy was identified as an independent risk factor for co-infections among both the COVID-19 and Influenza patients (OR 6.07 (1.15–35.73), $p = 0.033$, and 9.87 (1.54–197.90), $p = 0.047$, respectively).

Table 3. Univariate analysis of the risk factors for co-infections.

	COVID-19 n = 57		Influenza n = 55	
	OR (CI95%)	p-Value	OR (CI95%)	p-Value
Sex	0.53 (0.14–2.07)	0.347	1.07 (0.32–3.54)	0.912
Age (years)	0.99 (0.94–1.04)	0.597	0.94 (0.89–0.99)	0.026
Current smokers	1.82 (0.23–10.68)	0.521	2.49 (0.72–10.2)	0.17
Chronic pulmonary disease	1.06 (0.25–3.92)	0.932	0.79 (0.24–2.61)	0.7
Obesity (BMI $\geq$ 30 kg/m^2)	1.21 (0.32–4.29)	0.772	0.45 (0.12–1.65)	0.219
Arterial hypertension	0.54 (0.15–1.97)	0.345	0.31 (0.06–1.14)	0.099
Diabetes	*	0.053	0.33 (0.09–1.17)	0.085
Charlson Comorbidity index	1.01 (0.79–1.26)	0.964	0.66 (0.44–0.92)	0.022
Immunosuppressive therapy	6.07 (1.15–35.73)	0.033	6.74 (1.15–128.77)	0.08
Solid organ transplant	1.75 (0.08–19.85)	0.659	2.00 (0.29–40.14)	0.543
Baseline SOFA	1.16 (1–1.38)	0.066	1.01 (0.88–1.18)	0.866
Baseline SAPS 3	1 (0.97–1.04)	0.878	1.00 (0.96–1.04)	0.856
Baseline PaO_2/FiO_2 ratio	1 (0.99–1.01)	0.749	1.00 (0.99–1)	0.153
Influenza			9.03 (3.94–22.02)	<0.001

Univariate r UnU. Univariate regression analysis, except *: Fisher's exact test. A $p < 0.05$ was considered statistically significant. The data are presented as the odds ratio (OR) with its 95% confidence interval (CI 95%). Abbreviations—BMI: body mass index, *SOFA*: Sequential Organ Failure Assessment, SAPS 3: Simplified Acute Physiology Score 3, PaO_2/FiO_2 ratio: ratio of partial oxygen pressure to the fraction inspired air.

2.4. Secondary Infections

The secondary infections are presented in Table 4. The COVID-19 patients had significantly more secondary infections than the Influenza patients. The secondary infections were mostly caused by Gram-negative bacilli in both groups. Significantly more secondary infections were caused by multi-resistant bacteria in the COVID-19 group than in the Influenza group. On the other hand, the Influenza patients presented more secondary infections due to *Aspergillus* sp. than the COVID-19 patients (9/29, 31% vs. 1/60, 2%, $p < 0.001$). No viral secondary infections were reported for the first secondary infection in either cohort.

Table 4. Secondary infections.

	COVID-19	Influenza	*p*-Value
	N = 57	N = 55	
Total number of infectious events	60	29	-
Secondary infections			
Event 1, n (%)	37 (65)	16 (29)	<0.001
Time to onset (day)	8 (2–23)	6.5 (0–17)	0.484
Bacteremia, n (%) *	13 (35)	4 (25)	0.538
Respiratory infections, n (%) *	35 (95)	13 (81)	0.155
Others, n (%) *	3 (8)	3 (19)	0.351
Types of pathogens **			
Gram-positive cocci, n (%)	6 (16)	2 (20)	>0.999
Gram-negative bacilli, n (%)	34 (92)	9 (90)	>0.999
Virus, n (%)	0 (0)	0 (0)	>0.999
Aspergillus sp., n (%)	1 (3)	7 (44)	<0.001
Event 2, n (%)	18 (32)	7(13)	0.017
Time to onset (day)	18 (11–29)	16 (8–21)	0.048
Bacteremia, n (%) *	4 (22)	2 (29)	>0.999
Respiratory infections, n (%) *	13 (72)	5 (71)	>0.999
Others, n (%) *	2 (11)	0 (0)	>0.999
Types of pathogens **			
Gram-positive cocci, n (%)	3 (18)	0 (0)	>0.999
Gram-negative bacilli, n (%)	16 (94)	4 (100)	>0.999
Virus, n (%)	1 (6)	1 (14)	0.49
Aspergillus sp., n (%)	0 (0)	0 (0)	>0.999
Other fungi, n (%)	0 (0)	2 (40)	0.039
Event 3, n (%)	5 (9)	6 (11)	0.704
Time to onset (day)	23 (18–31)	22 (15–69)	0.583
Bacteremia, n (%) *	5 (100)	1 (33)	0.061
Respiratory infections, n (%) *	2 (40)	5 (83)	0.242
Others, n (%) *	0 (0)	0 (0)	>0.999
Types of pathogens **			
Gram-positive cocci, n (%)	2 (50)	1 (20)	0.524
Gram-negative bacilli, n (%)	4 (100)	5 (100)	>0.999
Virus, n (%)	1 (20)	3 (50)	0.546
Aspergillus sp., n (%)	0 (0)	2 (33)	0.456
Other fungi, n (%)	0 (0)	2 (33)	0.456

Continuous variables are reported as the median (interquartile range) and categorical variables are reported as numbers (percentages). Time is reported as the median (minimum–maximum). A $p < 0.05$ was considered statistically significant. * Secondary infections could be multisite. ** Secondary infections were sometimes polymicrobial.

The univariate logistic regression model of the risk factors for secondary infections are reported in Table 5. Table S2 provides the frequencies of each variable in the patients with secondary infections, compared to those without, in both cohorts of patients. Obesity, baseline SOFA scores, and treatment with vasopressors were identified as risk factors for secondary infections in the COVID-19 group, and support with Extra Corporeal Membrane Oxygenation (ECMO) was identified as such in both the COVID-19 and Influenza patients. Each variable for which the univariate OR yielded a p-value ≤ 0.1 was included in a multivariable logistic regression model. Treatment with vasopressors was the only independent risk factor identified for secondary infections in the COVID-19 group, with an OR of 16.23 (3.36–100.42; $p < 0.001$) and ECMO was the only independent risk factor identified for secondary infections in the Influenza group, with an OR of 22.8 (3.38–457.75; $p = 0.006$).

Table 5. Univariate analysis of the risk factors for secondary infections.

	COVID-19 n = 57		Influenza n = 55	
	OR (CI95%)	***p*-Value**	**OR (CI95%)**	***p*-Value**
Sex	2.42 (0.73–8.11)	0.146	1.1 (0.34–3.65)	0.871
Age (years)	0.98 (0.94–1.03)	0.472	0.97 (0.93–1.01)	0.139
Current smokers	1.09 (0.19–8.42)	0.924	1.12 (0.34–3.63)	0.852
Chronic pulmonary disease	0.99 (0.31–3.39)	0.983	0.51 (0.15–1.66)	0.274
Obesity	4.82 (1.33–23.19)	0.026	0.82 (0.2–2.96)	0.768
Arterial hypertension	1.39 (0.44–4.31)	0.568	0.72 (0.22–2.41)	0.587
Diabetes	*	0.548	0.46 (0.09–1.75)	0.286
Charlson Comorbidity Index	0.87 (0.7–1.07)	0.196	0.85 (0.61–1.17)	0.32
Immunosuppressive therapy	1.41 (0.27–10.52)	0.701	1.52 (0.39–5.46)	0.529
Solid organ transplant	*	0.545	0.45 (0.02–3.14)	0.487
Baseline SOFA	1.21 (1.04–1.45)	0.018	1.01 (0.86–1.17)	0.922
Baseline SAPS 3	0.99 (0.96–1.02)	0.482	1.03 (0.99–1.07)	0.192
Baseline PaO_2/FiO_2 ratio	0.99 (0.98–0.99)	0.004	1 (0.99–1)	0.425
Vasopressors	19.43 (4.8–104.01)	<0.001	3.92 (0.91–27.31)	0.098
ECMO	*	0.005	22.8 (3.38–457.75)	0.006
Co-infections	0.54 (0.15–1.97)	0.345	3.5 (0.81–24.45)	0.131
Influenza			0.22 (0.1–0.48)	<0.001

Univariate regression analysis, except *: Fisher's exact test. A $p < 0.05$ was considered statistically significant. The data are presented as the odds ratio (OR) with its 95% confidence interval (CI 95%). Abbreviations—BMI: body mass index, SOFA: Sequential Organ Failure Assessment, SAPS 3: Simplified Acute Physiology Score 3, PaO_2/FiO_2 ratio: ratio of partial oxygen pressure to the fraction inspired air, ECMO: extracorporeal membrane oxygenation.

2.5. Outcome and Risk Factors for ICU Death

The data on the outcomes are reported in Table 6. The COVID-19 patients stayed significantly longer in the ICU and needed non-significantly longer mechanical ventilation than the Influenza patients. ICU mortality was comparable between both groups, with an overall mortality of 43/112 (38%). The risk factors for ICU death were looked for in both cohorts. The factors explored were: sex, age, comorbidities, the Charlson Comorbidity Index, baseline SOFA, baseline SAPS 3, baseline PaO_2/FiO_2 ratio upon admission, the use of vasopressors, ECMO, co-infections, secondary infections, the number of secondary infectious events, and infections due to multi-drug resistant bacteria. In the COVID-19 group, obesity, hypertension, diabetes, the Charlson Comorbidity Index, and having a solid organ transplant were identified as risk factors for ICU death in the univariate analysis (Table 7). In both groups, baseline SOFA, SAPS3, and treatment with vasopressors were also identified as risk factors for ICU death. In the multivariable analysis, obesity, arterial hypertension, a high Charlson Comorbidity Index, and treatment with vasopressors were identified as independent risk factors for ICU death in the COVID-19 group, with ORs of 4.71 (1.07–23.54; $p= 0.44$), 4.97 (1.06–30.50; $p = 0.05$), 1.54 (1.13–2.29; $p = 0.014$), and 16.13 (2.02–377.47; $p = 0.25$), respectively. However, no risk factors for death were identified for the Influenza group. Furthermore, neither co-infections, secondary infections, nor the number of secondary infections were identified as risk factors for death in either the COVID-19 or Influenza groups.

Table 6. ICU Patient outcome.

	COVID-19	Influenza	*p*-Value
	N = 57	N = 55	
Supportive measures			
Optiflow, n (%)	8 (14)	15 (27)	0.083
NIV/CPAP, n (%)	32 (56)	31 (56)	0.981
Invasive mechanical ventilation, n (%)	41 (72)	36 (65)	0.591
Prone positioning, n (%)	35 (61)	9 (16)	<0.001
ECMO, n (%)	13 (23)	7 (13)	0.164
Vasopressors, n (%)	41 (72)	39 (71)	0.905
Renal replacement therapy, n (%)	16 (28)	9 (16)	0.137
ICU stay (days)	15 (4–29)	5 (2–10)	0.001
Mechanical ventilation duration (days)	11 (0–22)	4 (1–18)	0.716
ICU mortality (%), n (%)	20 (35)	23 (41)	0.464

Continuous variables are reported as the median (interquartile range), and categorical variables are reported as numbers (percentages). A $p < 0.05$ was considered statistically significant. Abbreviations—ICU: intensive care unit, NIV: non-invasive ventilation, CPAP: continuous positive airway pressure, ECMO: extracorporeal membrane oxygenation.

Table 7. Univariate analysis of the risk factors for ICU mortality.

	COVID-19 n = 57		Influenza n = 55	
	OR (CI95%)	*p*-Value	OR (CI95%)	*p*-Value
Sex	1.92 (0.556–7.81)	0.323	1.56 (0.529–0.72)	0.426
Age (years)	1.001 (0.764–1.05)	0.745	101 (0.975–1.05)	0.563
Current smokers	0.917 (0.119–5.18)	0.924	0.438 (0.136–1.32)	0.15
Chronic pulmonary disease	0.29 (0.06–1.06)	0.082	0.5 (0.164–1.47)	0.213
Obesity (BMI $\geq$ 30 kg/m^2)	5.44 (1.71–18.77)	0.005	0.958 (0.273–32)	0.945
Arterial hypertension	4.82 (1.33–23.19)	0.026	0.496 (0.16–1.49)	0.215
Diabetes	4.44 (1.15–19.48)	0.035	0.051 (0.003–0.293)	0.006
Charlson Comorbidity Index	1.37 (1.1–1.8)	0.012	0.858 (0.648–1.17)	0.264
Immunosuppressive therapy	2 (0.333–11.84)	0.425	0.533 (0.171–1.58)	0.264
Solid organ transplant		0.012 *	0.245 (0.012–1.67)	0.215
Baseline SOFA	1.22 (1.05–1.44)	0.012	1.18 (1.035–1.38)	0.019
Baseline SAPS	1.04 (1.001–1.09)	0.038	1.06 (1.02–1.11)	0.011
Baseline PaO_2/FiO_2 ratio	0.998 (0.991–1.004)	0.463	0.988 (0.977–0.996)	0.008
Vasopressors	12.95 (2.29–245.19)	0.018	4.56 (1.24–22.22)	0.034
ECMO	3.45 (0.936–13.62)	0.065	2.04 (0.406–11.3)	0.386
Co-infections	0.778 (0.187–2.87)	0.711	4 (1.08–19.55)	0.054
Secondary infections	3.05 (0.91–12.24)	0.087	1.60 (0.49–5.26)	0.432
Number of surinfections	1.415 (0.825–2.5)	0.211	1.14 (0.685–1.9)	0.606
Drug multiresistance bacteria	*	>0.999 *	0.463 (0.021–3.95)	0.519
Influenza			1.32 (0.62–2.87)	0.465

Univariate regression analysis, except *: Fisher's exact test. $p < 0.05$ was considered statistically significant. The data are presented as the odds ratio (OR) with its 95% confidence interval (CI95%). Abbreviations—BMI: body mass index, SOFA: Sequential Organ Failure Assessment, SAPS 3: Simplified Acute physiology Score 3, PaO_2/FiO_2 ratio: ratio of partial oxygen pressure to the fraction inspired air, ECMO: extracorporeal membrane oxygenation.

3. Discussion

In this study, we compared the co and secondary infections in critically ill COVID-19 patients admitted during the first wave of the pandemic in a single Belgian University hospital to those in critically ill Influenza patients. We also looked for risk factors for infectious events and death. To our knowledge, this is the first comparative study conducted specifically on ICU patients. We report fewer co-infections yet more secondary infections in the COVID-19 patients compared to the Influenza ICU patients. However, the time between the hospital arrival and the ICU admission was shorter—and the ICU length of stay was significantly longer—in the COVID-19 patients compared to the Influenza patients. Because our data come from the first pandemic wave and patient management has evolved since this moment, we will compare our results in this discussion to other first wave cohorts.

Although both COVID-19 and Influenza are viral diseases, the clinical presentations upon ICU admission differed between our two cohorts. At presentation, the COVID-19 patients in our cohort presented more ground glass features on a CT-scan and a lower leucocyte count, but they presented a higher lymphocyte count and higher c-reactive protein and lactate dehydrogenase values than the Influenza patients. These elements may help to differentiate between these diseases at the time of admission. Indeed, D'Onofrio et al. tried to identify factors that could help distinguish COVID-19 infection from Influenza infections in patients suspected of sepsis in the emergency department. They also reported a lower leucocyte count and higher lactate dehydrogenase values upon admission in the COVID-19 patients compared to the Influenza patients [17].

Despite similar Charlson scores upon ICU admission in our study, the Influenza patients had more comorbidities, and more of them were immunosuppressed compared to the COVID-19 patients. They were hospitalized for longer in the general ward before being admitted to the ICU, and they received more antibiotics before and during the first 48 h of ICU admission. This may partially explain why the number of co-infections observed among the Influenza group upon ICU admission was almost threefold higher than that observed among the COVID-19 cohort.

One out of five of our COVID-19 patients presented a co-infection, concordant with the prevalence rates of 4.3% to 28% previously reported in the literature [3–5,7,9]. However, if we consider only respiratory co-infections, two French studies reported higher rates than what we observed: 19.8% and 27.7% compared to 16%, respectively [5,8]. Half of the Influenza patients presented a bacterial co-infection, which was consistent with the values reported in the literature, varying from 5.9 to 51.1% [12].

The pathogens responsible for the co-infections were mostly bacterial, followed by viruses and fungi in both the Influenza and COVID-19 patients. We found 16% of co-infections with *Aspergillus* spp. in the Influenza group but none in the COVID-19 group. This observation could be due to the more frequent chronic consumption of corticosteroids in the Influenza group compared to the COVID-19 patients. Only IS therapy was identified as an independent risk factor by the multivariate analysis for co-infections in both groups. This risk factor for co-infections is consistent with the Influenza literature [13].

Secondary infections were observed more than two times as frequently in the COVID-19 patients than in the Influenza patients. Indeed, two out of three COVID-19 patients in our cohort presented at least one secondary infection, consistent with other studies on secondary infections in COVID-19 patients. Soriano et al., in a retrospective study, reported a secondary infections rate of 51% in 83 COVID-19 ICU patients [4]. Considering only ventilator-associated pneumonia (VAP), Razazi et al. reported that 64% of patients experienced at least one incidence of VAP within 8 (5–12) days of mechanical ventilation in a retrospective study on 90 patients [7]. This higher number of secondary infections may be partially explained by the longer ICU stay of the COVID-19 patients. The time to the onset of secondary infections was similar in the COVID-19 and Influenza cohorts, with a median time of 7 days. Bardi et al., who described nosocomial infections in COVID-19 ICU patients, found a time to onset of 9 days (IQR 5-11) [9]. Elabbadi et al., who described respiratory co and secondary bacterial infections in ICU patients, found a similar delay of 7.5 days [8]. Kokkoris et al., who described secondary blood stream infections in COVID-19 ICU patients, found a slightly longer time to onset of 11 days [10]. The rate of secondary infections may be more significant in the following waves of COVID-19, as glucocorticoids have become the standard of care [18]. Rothe and al. compared ICU COVID-19 patients treated or untreated with glucocorticoids (second versus first wave of COVID-19 in Germany). In their retrospective study, the use of dexamethasone was associated with more pulmonary infectious complications [19]. However, IS therapy was not identified as a risk factor for secondary infections in our study, possibly due to the small cohort size. Tocilizumab has also become the standard of care for COVID-19 patients, but it was not associated with more secondary infections in a recent meta-analysis [20].

The site of the secondary infections in the Influenza and COVID-19 patients was mostly respiratory, concordant with previous reports [4]. Thirty-five percent of these infections were bacteremia in the COVID-19 group, and the pathogens were mostly gram-negative bacilli, consistent with the literature (30.7–40%) [4,7,8,10]. As for co-infections, we found more secondary infections due to *Aspergillus* spp. in the Influenza group compared to the COVID-19 group. In the literature, Bardi et al. described two ventilator-associated infections due to *Aspergillus fumigatus* and one incidence of hospital-acquired pneumonia in COVID-19 ICU patients [9]. The only prospective multicentric study that evaluated coronavirus-associated pulmonary aspergillosis reported 27.7% of Aspergillosis-related secondary infections, with a median of 4 (2–8) days after ICU admission [21]. Various elements could explain this higher rate of infection compared to our observations: the study included only ARDS patients on mechanical ventilation (for more than 48 h), and most patients had received corticosteroids (60% in the aspergillus group and 46.6% in the non-aspergillus group).

The proportion of infections due to multi-drug resistant bacteria was 32% in the COVID-19 group, similar to the findings published by Bardi et al., who reported that 31% of the observed nosocomial infections in their cohort were due to multidrug resistant microorganisms [9]. The number of infections due to multidrug resistant bacteria was significantly greater in the COVID-19 patients than in the Influenza patients during the first secondary event, but it was similar for the second and third events. These results were not expected given the fact that the Influenza group consumed more antibiotics than the COVID-19 cohort at the time of ICU admission, and previous antibiotic therapy is a recognized risk factor for infections due to the multidrug resistant pathogens in the ICU setting [22].

The COVID-19 patients had a longer ICU stay and spent more time on mechanical ventilation than the Influenza patients, as previously described [15]. Despite these elements, the ICU mortality between the two groups in our cohort was comparable. On the other hand, Piroth et al., who compared COVID-19 patients with 2018–2019 seasonal Influenza patients, found a higher ICU mortality in the COVID-19 group (27.1%) than in the Influenza group (18%) [15]. In the literature, the ICU mortality of first wave COVID-19 patients varied from 24.1% to 36% [4,9]. In our study, neither co nor secondary infections were identified as risk factors for ICU death. Kreitman et al., who prospectively described bacterial respiratory co-infections in COVID-19 patients, also found no differences in terms of mortality between the co-infected and non-co-infected patients [6]. Our data are concordant with these results.

The major study limitations are the small sample size and the retrospective design. The small sample size may partially explain why we did not identify more risk factors for co and secondary infections. The rates of co and secondary infections in our study are concordant with the current literature. However, these rates may be overestimated, as the clinical and biological signs of co and secondary infection, such as fever and CRP, are also elevated in COVID-19 without infectious complications, making the differential diagnosis between simple colonization and infection difficult in this context. This study is nevertheless of interest because it analyzes microbiological events in COVID-19 patients before the introduction of other therapies such as glucocorticoids, convalescent plasma, tociluzimab, jak-inhibitors, etc., It also shows that there are differences between Influenza and COVID-19 in terms of disease presentation.

4. Materials and Methods

We performed a retrospective, monocentric study at Erasme Hospital, a 1048-bed teaching hospital (including 32 ICU beds) in Brussels, Belgium. We compared the infectious complications observed in the ICU between confirmed COVID-19 and confirmed flu patients from the 13th of March until the 20th of April 2020 and from the 1st of January 2015 until the 20th of April 2020, respectively.

We included all ICU patients with a positive Reverse Transcriptase Polymerase Chain Reaction (RT-PCR), a rapid antigen detection test (RDT), or a viral culture for Influenza A or B or SARS-CoV-2 on nasopharyngeal (NP) swabs or bronchoalveolar lavage (BAL) prior to or within 48 h of ICU admission. The exclusion criteria were patients younger than eighteen years old and pregnant women.

At our institution, Influenza is actively searched for during the yearly flu epidemic (based on Sciensano, the public health institute's epidemic curves) in all patients admitted for respiratory symptoms or fever using RDT for Influenza A and B (Influ A + B K-SeT, Coris bioConcept, Gembloux, Belgium), viral PCR (cobas® Influenza A/B and RSV Assay (Roche Molecular Systems, Pleasanton, CA, USA)), and cultures. Viral cultures are performed on confluent Vero (African green monkey kidney), MRC5 (human lung), and LLC-MK2 (rhesus monkey kidney) cells (Vircell, Santa-Fé, Spain) in 24-well or 6-well tissue culture plates (Greiner-Bio One, Frickenhausen, Germany) on all NP swabs performed for Influenza A and B detection and on all BAL samples. A multiplex PCR panel with Influenza A and B is also performed on all BAL specimens and NP swabs [23]. As there were no Influenza patients requiring intensive care at our institution during the first wave of the COVID-19 pandemic, we included ICU Influenza patients from the five previous Influenza seasons to obtain a cohort size similar to that of the cohort of ICU COVID-19 patients.

Furthermore, as of the 4th of March 2020, the beginning of the epidemic in Belgium, NP swabs for SARS-CoV-2 PCR were performed on all patients admitted to our hospital. Repeat SARS-CoV-2 PCRs were performed on NP swabs or BALs in patients with initial negative results for SARS-CoV-2 but with respiratory symptoms or an unexplained fever. Influenza was excluded by either an RDT test and culture on an NP swab or a PCR and culture on BAL in every SARS-CoV-2 suspected patient. The RT-PCR tests for SARS-CoV2 were performed using various commercialized automated PCR systems.

In the ICU, multiple biological samples for the microbiology laboratory are collected systematically from all patients admitted with, or who develop during their stay, clinical (i.e., hypotension, fever, cough, change in consciousness, etc.), biological (i.e., elevated C-reactive protein, elevated leucocyte count, etc.), or radiological signs of infection. The sampling, oriented by the clinical exam, includes NP swabs, sputum, BALs, blood, urine, liquid from surgical drains, swabs, or punctures of purulent lesions to perform direct microscopy and culture. We use a customized TaqMan® array card real-time PCR method, targeting 24 viruses, 8 bacteria, and 2 fungi simultaneously [23] on BAL, along with nasal swabs. The results are considered positive if the threshold cycle is below 35. At the start of this pandemic, because of the limited amount of reagent, we had to limit the multiplex panel to ICU patients who underwent a BAL for respiratory deterioration. On BAL, virus/bacteria, mycobacteria, and fungal cultures are systematically performed, as well as galactomannan detection. Galactomannan is also measured on the blood samples if there is clinical suspicion of invasive aspergillosis infection and systematically (twice a week) in immunocompromised patients at a high risk of developing invasive aspergillosis [24]. Furthermore, ICU patients undergo systematic (twice a week) microbiological sampling, including using nasal swabs to detect methicillin resistant *Staphylococcus aureus*, rectal swabs for the screening of multi-resistant Gram-negative bacteria carriage, and respiratory and drain or catheter samples. A PCR on blood for cytomegalovirus detection is also performed twice a week in immunocompromised patients. Every patient is discussed together with the ICU team, infectious diseases (ID) team, and microbiologists on a bi-weekly basis, completed with further daily consults by the ID specialist, if needed, to decide on the best anti-infectious management.

When patients are hospitalized in the ICU with ARDS, respiratory deterioration is systematically documented with a thoracic CT-scan to distinguish secondary infections from other respiratory complications (i.e., pulmonary embolism, pneumothorax, etc.), along with a BAL (if the patient is intubated), tracheal aspiration, or sputum sample.

4.1. Data Collection

The medical records were reviewed by two medical doctors. The data collection included each patient's demographics, comorbidities, reason for ICU admission, delay between hospital and ICU admission, diagnosis upon ICU admission, clinical, microbiological, and laboratory data, radiological results, and outcome. For demographics, age and sex were recorded. For the comorbidities, we calculated the Charlson Comorbidity Index [25] and recorded the presence of chronic pulmonary disease, IS therapy, solid organ transplantation, cardiovascular disease, hypertension, obesity (defined as a body mass index (BMI) greater than or equal to 30 kg/m^2), neoplasia, active smoking, diabetes, and renal insufficiency (defined as a glomerular filtration rate below 60 mL/min using the CKD–EPI equation [26]). For clinical data, the Simplified Acute Physiology score 3 (SAPS 3) [27] and the PaO_2/FiO_2 ratio upon admission, along with the Sepsis-related Organ Failure Assessment (SOFA) score [28] during the first 24 h in the ICU, were recorded, as well as the need for supplementary oxygen, mechanical ventilation, ventral decubitus, catecholamines, and ECMO during the ICU stay. For the laboratory analyses (d-dimer, total leucocyte, lymphocyte and platelet counts, creatinine, bilirubin, C-reactive protein, lactate dehydrogenase, and creatinine kinase values), the radiological findings (thoracic CT-scan findings: pulmonary embolism, condensations, micronodules, or round glass) were recorded upon admission and at every clinical deterioration.

We considered a documented infection, the association of clinical symptoms (depending on the site of infection), biological signs of infection (see above), and the presence of a relevant pathogen on the microbiological samples. The presence of pathogens had to be correlated with the decision to treat the event—unless the pathogen was a virus (as the infection was considered to be the cause of the clinical deterioration). The relevant pathogens were defined as described below.

For blood cultures (BC), coagulase-negative *Staphylococci* and *Corynebacterium* spp. were considered relevant only if they were isolated in more than one bottle with similar antibiotic susceptibility profiles and persistent after catheter removal/change. All other pathogens identified in the blood cultures were considered relevant. For urinary tract infections, the patients had to have symptoms or fever, an elevated urine white blood cells count, and a positive urine culture (>10^5 colony forming units/mL) of no more than two isolated micro-organisms. For the purulent lesions obtained by puncture, all of the pathogens identified by the culture were considered relevant. For surgical drain cultures, only the pathogens identified in the samples taken within the first 24 h of placing the drain were considered relevant. For respiratory samples, we excluded the pathogens belonging to the mouth microbiota. As proposed by Verweij et al., an invasive pulmonary Aspergillosis infection was defined as the presence of pulmonary infiltrates with at least one of the following: a galactomannan index superior to one (on blood or BAL) or a positive BAL culture for *Aspergillus* sp. [29].

An antibiogram is systematically performed on relevant bacterial pathogens. Multi-resistant bacteria were defined as having acquired non-susceptibility to at least one agent in three or more antibiotic classes [4]. If an infection was polymicrobial, even if only one of the bacteria was multi-resistant (as defined above), the patient was considered infected by a multi-resistant pathogen.

The infections were defined as co or secondary infections according to the timing of their diagnosis. Co-infections were those diagnosed from 48 h prior to admission until 48 h after admission. All episodes occurring after the first 48 h of admission were recorded as secondary infections. For each infectious episode, the site, the pathogen(s) responsible for the infection, their resistance profile, and the delay between the admission and the event were recorded. The anti-infectious treatments administered were recorded with a distinction between anti-infectious therapies given before ICU admission, those administered empirically during the first 48 h of ICU admission, and those administered after the first microbiological results. Subsequent anti-infectious therapy given for subsequent infectious episodes was also recorded. For the outcome, the length of the ICU-stay, the 28-day mortality, the ICU, and the hospital mortality were recorded.

4.2. Statistics

The discrete variables were presented as numbers (%) and compared using the Chi-square-test or Fisher's exact test, and the continuous variables were expressed as the median and range if not normally distributed and compared with the Mann–Whitney U-test. A univariate, followed by a multivariate logistic regression model, were built to determine the independent risks factors for co-infections, secondary infections, and ICU deaths for the COVID-19 and the Influenza cohorts. The independent risk factors were identified after multivariate analyses models were derived from a backward stepwise analysis from a "full logistic regression model" including all variables for which the univariate odds ratio (OR) yielded a p-value ≤ 0.1. A p-value < 0.05 was considered statistically significant. The statistical software R (version 4.0.3) [30] was used.

5. Conclusions

In conclusion, we report fewer co-infections yet more secondary infections among critically ill COVID-19 patients from the first wave of the pandemic compared to Influenza ICU patients. Most of the infectious complications were respiratory and of bacterial origin. Invasive aspergillosis infections were only observed in the Influenza patients. Immunosuppressive therapy was identified as an independent risk factor for co-infections among both COVID-19 and Influenza patients. Only ECMO was identified as an independent risk factor for secondary infections in the Influenza patients, and treatment with vasopressors was identified as an independent risk factor for secondary infections in the COVID-19 cohort. Finally, the mortality rate was not greater for the patients with co or secondary infections among both the Influenza and COVID-19 patients.

Our study provides a good picture of the natural course of the COVID-19 disease in critically ill patients. Today, glucocorticoids are the standard of care, and many patients receive anti-Interleukine-6 therapy [31]. Co-infections, secondary infections, and outcomes may differ significantly among the COVID-19 cohort described in this study. Further studies comparing today's critically ill COVID-19 patients to Influenza patients in terms of co-infections, secondary infections, and outcomes should be pursued. The risk factors for co and secondary infections in critically ill COVID-19 patients need to be better defined to guide antibiotherapy prescription for these patients. On the other hand, antibiotics should be given quickly to immunosuppressed Influenza patients admitted to the ICU, as bacterial co-infections are frequent in these patients.

Supplementary Materials: The following supporting information can be downloaded at: https://www.mdpi.com/article/10.3390/antibiotics11060704/s1, Table S1: Patients characteristics according to their co-infection status, Table S2: Patients characteristics according to their secondary infection status.

Author Contributions: Conceptualization, D.M., I.E. and M.H.; methodology, D.M., I.E., A.V.M. and M.H.; data curation, D.M., E.G.B., N.Y., A.V.M. and M.H.; writing—original draft preparation, D.M. and M.H.; writing—review and editing, D.M., E.G.B., N.Y., F.S.T. and M.H.; supervision, M.H. All authors have read and agreed to the published version of the manuscript.

Funding: This research received no external funding.

Institutional Review Board Statement: This study was conducted in accordance with the Declaration of Helsinki and approved by the Ethics Committee of Erasme (protocol code P2020/258 on the 4th of May 2020).

Informed Consent Statement: Patient consent was waived due to the retrospective design and anonymous data analysis and publication.

Data Availability Statement: The data presented in this study are available on request from the corresponding author. The data are not publicly available due to the medical nature of this data, even if the data is anonymized.

Conflicts of Interest: The authors declare no conflict of interest.

References

1. Mortality Analyses. Available online: https://coronavirus.jhu.edu/data/mortality (accessed on 22 August 2021).
2. Lescure, F.-X.; Bouadma, L.; Nguyen, D.; Parisey, M.; Wicky, P.-H.; Behillil, S.; Gaymard, A.; Bouscambert-Duchamp, M.; Donati, F.; Hingrat, Q.L.; et al. Clinical and Virological Data of the First Cases of COVID-19 in Europe: A Case Series. *Lancet Infect. Dis.* **2020**, *20*, 697–706. [CrossRef]
3. Lansbury, L.; Lim, B.; Baskaran, V.; Lim, W.S. Co-Infections in People with COVID-19: A Systematic Review and Meta-Analysis. *J. Infect.* **2020**, *81*, 255–275. [CrossRef] [PubMed]
4. Soriano, M.C.; Vaquero, C.; Ortiz-Fernández, A.; Caballero, A.; Blandino-Ortiz, A.; de Pablo, R. Low Incidence of Co-Infection, but High Incidence of ICU-Acquired Infections in Critically Ill Patients with COVID-19. *J. Infect.* **2020**, *82*, e20–e21. [CrossRef] [PubMed]
5. Contou, D.; Claudinon, A.; Pajot, O.; Micaëlo, M.; Longuet Flandre, P.; Dubert, M.; Cally, R.; Logre, E.; Fraissé, M.; Mentec, H.; et al. Bacterial and Viral Co-Infections in Patients with Severe SARS-CoV-2 Pneumonia Admitted to a French ICU. *Ann. Intensive Care* **2020**, *10*, 119. [CrossRef]
6. Kreitmann, L.; Monard, C.; Dauwalder, O.; Simon, M.; Argaud, L. Early Bacterial Co-Infection in ARDS Related to COVID-19. *Intensive Care Med.* **2020**, *46*, 1787–1789. [CrossRef]
7. Razazi, K.; Arrestier, R.; Haudebourg, A.F.; Benelli, B.; Carteaux, G.; Decousser, J.-W.; Fourati, S.; Woerther, P.L.; Schlemmer, F.; Charles-Nelson, A.; et al. Risks of Ventilator-Associated Pneumonia and Invasive Pulmonary Aspergillosis in Patients with Viral Acute Respiratory Distress Syndrome Related or Not to Coronavirus 19 Disease. *Crit. Care Lond. Engl.* **2020**, *24*, 699. [CrossRef]
8. Elabbadi, A.; Turpin, M.; Gerotziafas, G.T.; Teulier, M.; Voiriot, G.; Fartoukh, M. Bacterial Coinfection in Critically Ill COVID-19 Patients with Severe Pneumonia. *Infection* **2021**, *49*, 559–562. [CrossRef]
9. Bardi, T.; Pintado, V.; Gomez-Rojo, M.; Escudero-Sanchez, R.; Azzam Lopez, A.; Diez-Remesal, Y.; Martinez Castro, N.; Ruiz-Garbajosa, P.; Pestaña, D. Nosocomial Infections Associated to COVID-19 in the Intensive Care Unit: Clinical Characteristics and Outcome. *Eur. J. Clin. Microbiol. Infect. Dis. Off. Publ. Eur. Soc. Clin. Microbiol.* **2021**, *40*, 495–502. [CrossRef]
10. Kokkoris, S.; Papachatzakis, I.; Gavrielatou, E.; Ntaidou, T.; Ischaki, E.; Malachias, S.; Vrettou, C.; Nichlos, C.; Kanavou, A.; Zervakis, D.; et al. ICU-Acquired Bloodstream Infections in Critically Ill Patients with COVID-19. *J. Hosp. Infect.* **2021**, *107*, 95–97. [CrossRef]
11. Langford, B.J.; So, M.; Raybardhan, S.; Leung, V.; Westwood, D.; MacFadden, D.R.; Soucy, J.-P.R.; Daneman, N. Bacterial Co-Infection and Secondary Infection in Patients with COVID-19: A Living Rapid Review and Meta-Analysis. *Clin. Microbiol. Infect. Off. Publ. Eur. Soc. Clin. Microbiol. Infect. Dis.* **2020**, *26*, 1622–1629. [CrossRef]
12. Klein, E.Y.; Monteforte, B.; Gupta, A.; Jiang, W.; May, L.; Hsieh, Y.-H.; Dugas, A. The Frequency of Influenza and Bacterial Coinfection: A Systematic Review and Meta-Analysis. *Influenza Other Respir. Viruses* **2016**, *10*, 394–403. [CrossRef] [PubMed]
13. Martin-Loeches, I.; Schultz, M.J.; Vincent, J.-L.; Alvarez-Lerma, F.; Bos, L.D.; Solé-Violán, J.; Torres, A.; Rodriguez, A. Increased Incidence of Co-Infection in Critically Ill Patients with Influenza. *Intensive Care Med.* **2017**, *43*, 48–58. [CrossRef] [PubMed]
14. McCullers, J.A. The Co-Pathogenesis of Influenza Viruses with Bacteria in the Lung. *Nat. Rev. Microbiol.* **2014**, *12*, 252–262. [CrossRef] [PubMed]
15. Piroth, L.; Cottenet, J.; Mariet, A.-S.; Bonniaud, P.; Blot, M.; Tubert-Bitter, P.; Quantin, C. Comparison of the Characteristics, Morbidity, and Mortality of COVID-19 and Seasonal Influenza: A Nationwide, Population-Based Retrospective Cohort Study. *Lancet Respir. Med.* **2021**, *9*, 251–259. [CrossRef]
16. Shafran, N.; Shafran, I.; Ben-Zvi, H.; Sofer, S.; Sheena, L.; Krause, I.; Shlomai, A.; Goldberg, E.; Sklan, E.H. Secondary Bacterial Infection in COVID-19 Patients Is a Stronger Predictor for Death Compared to Influenza Patients. *Sci. Rep.* **2021**, *11*, 12703. [CrossRef]
17. D'Onofrio, V.; Van Steenkiste, E.; Meersman, A.; Waumans, L.; Cartuyvels, R.; Van Halem, K.; Messiaen, P.; Gyssens, I.C. Differentiating Influenza from COVID-19 in Patients Presenting with Suspected Sepsis. *Eur. J. Clin. Microbiol. Infect. Dis. Off. Publ. Eur. Soc. Clin. Microbiol.* **2021**, *40*, 987–995. [CrossRef]
18. COVID-19 Treatment Guidelines Panel. Coronavirus Disease 2019 (COVID-19) Treatment Guidelines. National Institutes of Health. Available online: https://Www.Covid19treatmentguidelines.Nih.Gov/ (accessed on 17 August 2021).
19. Rothe, K.; Lahmer, T.; Rasch, S.; Schneider, J.; Spinner, C.D.; Wallnöfer, F.; Wurst, M.; Schmid, R.M.; Waschulzik, B.; Fuest, K.; et al. Dexamethasone Therapy and Rates of Secondary Pulmonary and Bloodstream Infections in Critically Ill COVID-19 Patients. *Multidiscip. Respir. Med.* **2021**, *16*, 793. [CrossRef]
20. Rubio-Rivas, M.; Forero, C.G.; Mora-Luján, J.M.; Montero, A.; Formiga, F.; Homs, N.A.; Albà-Albalate, J.; Sánchez, L.; Rello, J.; Corbella, X. Beneficial and Harmful Outcomes of Tocilizumab in Severe COVID-19: A Systematic Review and Meta-analysis. *Pharmacotherapy* **2021**, *41*, 884–906. [CrossRef]

21. Bartoletti, M.; Pascale, R.; Cricca, M.; Rinaldi, M.; Maccaro, A.; Bussini, L.; Fornaro, G.; Tonetti, T.; Pizzilli, G.; Francalanci, E.; et al. Epidemiology of Invasive Pulmonary Aspergillosis Among Intubated Patients With COVID-19: A Prospective Study. *Clin. Infect. Dis.* **2020**, *73*, e3606–e3614. [CrossRef]
22. Timsit, J.-F.; Bassetti, M.; Cremer, O.; Daikos, G.; de Waele, J.; Kallil, A.; Kipnis, E.; Kollef, M.; Laupland, K.; Paiva, J.-A.; et al. Rationalizing Antimicrobial Therapy in the ICU: A Narrative Review. *Intensive Care Med.* **2019**, *45*, 172–189. [CrossRef]
23. Steensels, D.; Reynders, M.; Descheemaeker, P.; Curran, M.D.; Jacobs, F.; Denis, O.; Delforge, M.-L.; Montesinos, I. Clinical Evaluation of a Multi-Parameter Customized Respiratory TaqMan(®) Array Card Compared to Conventional Methods in Immunocompromised Patients. *J. Clin. Virol. Off. Publ. Pan Am. Soc. Clin. Virol.* **2015**, *72*, 36–41. [CrossRef] [PubMed]
24. Patterson, T.F.; Thompson, G.R.; Denning, D.W.; Fishman, J.A.; Hadley, S.; Herbrecht, R.; Kontoyiannis, D.P.; Marr, K.A.; Morrison, V.A.; Nguyen, M.H.; et al. Practice Guidelines for the Diagnosis and Management of Aspergillosis: 2016 Update by the Infectious Diseases Society of America. *Clin. Infect. Dis. Off. Publ. Infect. Dis. Soc. Am.* **2016**, *63*, e1–e60. [CrossRef] [PubMed]
25. Quan, H.; Li, B.; Couris, C.M.; Fushimi, K.; Graham, P.; Hider, P.; Januel, J.-M.; Sundararajan, V. Updating and Validating the Charlson Comorbidity Index and Score for Risk Adjustment in Hospital Discharge Abstracts Using Data from 6 Countries. *Am. J. Epidemiol.* **2011**, *173*, 676–682. [CrossRef]
26. Levey, A.S.; Stevens, L.A.; Schmid, C.H.; Zhang, Y.; Castro, A.F.; Feldman, H.I.; Kusek, J.W.; Eggers, P.; Van Lente, F.; Greene, T.; et al. A New Equation to Estimate Glomerular Filtration Rate. *Ann. Intern. Med.* **2009**, *150*, 604–612. [CrossRef]
27. Moreno, R.P.; Metnitz, P.G.H.; Almeida, E.; Jordan, B.; Bauer, P.; Campos, R.A.; Iapichino, G.; Edbrooke, D.; Capuzzo, M.; Le Gall, J.-R. SAPS 3—From Evaluation of the Patient to Evaluation of the Intensive Care Unit. Part 2: Development of a Prognostic Model for Hospital Mortality at ICU Admission. *Intensive Care Med.* **2005**, *31*, 1345–1355. [CrossRef] [PubMed]
28. Ferreira, F.L.; Bota, D.P.; Bross, A.; Mélot, C.; Vincent, J.L. Serial Evaluation of the SOFA Score to Predict Outcome in Critically Ill Patients. *JAMA* **2001**, *286*, 1754–1758. [CrossRef]
29. Verweij, P.E.; Rijnders, B.J.A.; Brüggemann, R.J.M.; Azoulay, E.; Bassetti, M.; Blot, S.; Calandra, T.; Clancy, C.J.; Cornely, O.A.; Chiller, T.; et al. Review of Influenza-Associated Pulmonary Aspergillosis in ICU Patients and Proposal for a Case Definition: An Expert Opinion. *Intensive Care Med.* **2020**, *46*, 1524–1535. [CrossRef]
30. R Core Team. *R A Language and Environment for Statistical Computing*; R Foundation for Statistical Computing: Vienna, Austria, References-Scientific Research Publishing; 2019. Available online: https://www.scirp.org/%28S%28i43dyn45teexjx455qlt3d2q%29%29/reference/referencespapers.aspx?referenceid=2631126 (accessed on 19 August 2021).
31. Therapeutics and COVID-19: Living Guideline. Available online: https://www.who.int/publications-detail-redirect/WHO-2019-nCoV-therapeutics-2021.3 (accessed on 9 January 2022).

antibiotics

MDPI

Article

Antimicrobial Use in Hospitalised Patients with COVID-19: An International Multicentre Point-Prevalence Study

Lea Papst [1,2,*], Roberto Luzzati [3], Biljana Carević [4], Carlo Tascini [5], Nina Gorišek Miksić [6], Vera Vlahović Palčevski [7], Zorana M. Djordjevic [8], Omar Simonetti [3], Emanuela Sozio [5], Milica Lukić [1,2], Goran Stevanović [4], Davor Petek [6] and Bojana Beović [1,2,†] on behalf of the COVID-PPS Study Group

1 Department of Infectious Diseases, University Medical Centre Ljubljana, Zaloška cesta 2, 1000 Ljubljana, Slovenia; milica.lukic@kclj.si (M.L.); bojana.beovic@kclj.si (B.B.)
2 Faculty of Medicine, University of Ljubljana, Vrazov trg 2, 1000 Ljubljana, Slovenia
3 Department of Infectious Diseases, Azienda Sanitaria Universitaria Giuliano Isontina, Via Giacomo Puccini 50, 34148 Trieste, Italy; roberto.luzzati@asugi.sanita.fvg.it (R.L.); omar.simonetti@asugi.sanita.fvg.it (O.S.)
4 Department of Hospital Epidemiology, University Clinical Centre of Serbia, Pasterova 2, 11000 Belgrade, Serbia; biljana.carevic@gmail.com (B.C.); goran_drste@yahoo.com (G.S.)
5 Infectious Diseases Clinic, Azienda Sanitaria Universitaria Friuli Centrale, Via Pozzuolo 33, 33100 Udine, Italy; carlo.tascini@asufc.sanita.fvg.it (C.T.); emanuela.sozio@gmail.com (E.S.)
6 Department of Infectious Diseases, University Medical Centre Maribor, Ljubljanska ulica 5, 2000 Maribor, Slovenia; nina.gorisekmiksic@ukc-mb.si (N.G.M.); davor.petek@ukc-mb.si (D.P.)
7 Unit of Clinical Pharmacology, Clinical Hospital Center Rijeka, Krešimirova ulica 42, 51000 Rijeka, Croatia; vera.vlahovicpalcevski@gmail.com
8 Department of Hospital Infections Control, University Clinical Centre Kragujevac, Zmaj Jovina 30, 34000 Kragujevac, Serbia; drzorana.25@gmail.com
* Correspondence: lea.papst@kclj.si
† Membership of the Group Name is provided in the Acknowledgments.

Citation: Papst, L.; Luzzati, R.; Carević, B.; Tascini, C.; Gorišek Miksić, N.; Vlahović Palčevski, V.; Djordjevic, Z.M.; Simonetti, O.; Sozio, E.; Lukić, M.; et al. Antimicrobial Use in Hospitalised Patients with COVID-19: An International Multicentre Point-Prevalence Study. *Antibiotics* **2022**, *11*, 176. https://doi.org/10.3390/antibiotics11020176

Academic Editors: Sabine Danielle Allard and Johan Van Laethem

Received: 19 December 2021
Accepted: 26 January 2022
Published: 28 January 2022

Abstract: Studies suggest that the incidence of coinfections in patients with the coronavirus disease 2019 (COVID-19) is low, but a large number of patients receive antimicrobials during hospitalisation. This may fuel a rise in antimicrobial resistance (AMR). We conducted a multicentre point-prevalence survey in seven tertiary university hospitals (in medical wards and intensive care units) in Croatia, Italy, Serbia and Slovenia. Of 988 COVID-19 patients, 521 were receiving antibiotics and/or antifungals (52.7%; range across hospitals: 32.9–85.6%) on the day of the study. Differences between hospitals were statistically significant (χ^2 (6, N = 988) = 192.57, $p < 0.001$). The majority of patients received antibiotics and/or antifungals within 48 h of admission (323/521, 62%; range across hospitals: 17.4–100%), their most common use was empirical (79.4% of prescriptions), and pneumonia was the main indication for starting the treatment (three-quarters of prescriptions). The majority of antibiotics prescribed (69.9%) belonged to the "Watch" group of the World Health Organization AWaRe classification. The pattern of antimicrobial use differed across hospitals. The data show that early empiric use of broad-spectrum antibiotics is common in COVID-19 patients, and that the pattern of antimicrobial use varies across hospitals. Judicious use of antimicrobials is warranted to prevent an increase in AMR.

Keywords: COVID-19; antimicrobial use; multicentre; point-prevalence study

1. Introduction

Antimicrobial resistance (AMR) is an inevitable consequence of antimicrobial use, making antimicrobial stewardship (AMS) an irreplaceable tool in the fight against increasing resistance [1]. Amidst the ever-increasing threat of AMR, the coronavirus disease 2019 (COVID-19) pandemic has shifted the focus of healthcare providers to the care of patients infected with severe acute respiratory syndrome coronavirus 2 (SARS-CoV-2). Since the pandemic began in 2020, nearly 270 million cases of COVID-19 have been confirmed, claiming more than 5.3 million lives worldwide [2].

Studies of bacterial and fungal infections in patients with COVID-19 suggest that the incidence of these infections is low, but a large number of patients receives antibiotics upon admission to hospital or during hospitalisation. Early bacterial coinfection has been reported in only 1.2–3.5% of patients with COVID-19 [3–6]. In rapid systematic reviews, bacterial or fungal infections have been identified in 6.9–8% of patients [7,8], with early bacterial coinfection being rare (3.5% of patients), and secondary bacterial/fungal infection occurring in 14.3% of patients [8]. Studies and systematic reviews reported the use of antimicrobials in up to 38.3–74.6% of patients [4,5,7,9,10].

It is still unclear how the pandemic and antibiotic use in patients with COVID-19 will affect AMR, as evidence-based data on AMR is still lacking [11]. Some studies have reported an increase in infections with multi-drug resistant microorganisms (MDR) [12–15], while others have found no increase in infections with MDR bacteria and fungi [6,16]. The differences suggest heterogeneous antibiotic use and infection control measures in COVID-19 patients in hospitals.

To better guide antimicrobial prescribing and adjust AMS programmes, more information is needed on antimicrobial use in patients with COVID-19 from different parts of the world. We conducted an international multicentre point-prevalence survey to collect comprehensive data on the characteristics and differences of antibiotic and antifungal prescribing in hospitalised patients with COVID-19 in various countries and hospitals in South-Eastern Europe.

2. Results

Of 988 COVID-19 patients, 521 were receiving antibiotics and/or antifungals on the day of the study (52.7%; range across hospitals: 32.9–85.6%). Differences between hospitals were statistically significant (χ^2 (6, N = 988) = 192.57, $p < 0.001$). Use of antibiotics and antifungals was common in intensive care units (ICUs) (135/186, 72.6%; range across hospitals: 54.1–100%) as well as in medical wards (386/802, 48.1%; range across hospitals: 14.3–93.6%). The characteristics of patients who received antibiotic or antifungal therapy are shown in Table 1. The majority of patients were male (61.2%), with an even higher proportion of male patients in ICUs (73.3%). The median age was 69 years, and the most common chronic disease was arterial hypertension. Approximately 75% of patients on antibiotics and/or antifungals were receiving concomitant corticosteroids and supplemental oxygen. Approximately two-thirds of patients were receiving one antibiotic/antifungal, others two or more. The median levels of C-reactive protein (CRP) and procalcitonin (PCT) at the start of antimicrobial therapy were 86.7 mg/L and 0.2 µg/L, respectively. For chronic diseases as well as for CRP and WBC (white blood count), statistically significant differences between hospitals were observed (Table 1). In ICU, 71.9% of patients were intubated and mechanically ventilated (range across hospitals: 20–95%), 14.1% were pronated (range across hospitals: 0–50%), 40% were receiving vasoactive support (range across hospitals: 13.3–70%) and 0.74% were on extracorporeal membrane oxygenation (ECMO) (range across hospitals: 0–3.4%).

The majority of patients received an antimicrobial agent within 48 h of admission (323/521, 62%; range across hospitals: 17.4–100%), 70.2% in medical wards (271/386; range across hospitals: 23.5–100%) and 38.5% in ICUs (52/135; range across hospitals: 0–100%). A total of 743 antibiotics and antifungals was prescribed. Their use was most commonly empirical (79.4% of prescriptions) (Table 2). Targeted treatment was more common in ICUs, especially with antibiotics and antifungals started later in the course of hospitalisation ($\geq$48 h after admission); 47% in ICUs vs. 31.3% in medical wards. Antibiotics and/or antifungals were prescribed for pneumonia in three-quarters of cases, followed by urinary tract infections (5.5% of prescriptions) and infections of no known origin (5.3% of prescriptions) (Table 3). While pneumonia was almost the only indication for antibiotics and/or antifungals being prescribed at the beginning of hospitalisation (88%

of prescriptions), less than half of them were prescribed due to pneumonia in medical wards later in the hospitalisation (45.2%), with other hospital-acquired infections gaining importance. In ICUs pneumonia remained the main indication for treatment with antibiotics and/or antifungals (75.5% of prescriptions). According to the treating physicians, the reasons for starting antimicrobial therapy were clinical presentation in 85.4% of cases, laboratory findings in 72.2% and imaging in 62.6% of cases.

Table 1. Characteristics of patients receiving antibiotics and/or antifungals.

	Total (N = 521) % (Range across Hospitals, %)	**Differences between Hospitals**		**Medical Wards (N = 386) % (Range across Hospitals, %)**	**ICUs (N = 135) % (Range across Hospitals, %)**
		X^2/Fisher's Exact *t*-Test (6, N = 521)	**ANOVA**		
Sex		0.67, $p = 0.995$			
Male	61.2% (58.6–65.2%)			57% (44.4–59.1%)	73.3% (59.1–100%)
Female	38.8% (34.8–41.4%)			43% (40.9–55.6%)	26.7% (0–40.9%)
Age in years, median (IQR), (range across hospitals)	69 (17) (62–75)		F (6, 514) = 7.21, $p < 0.001$	69 (19) (61–79)	68 (12.5) (62.5–75.5)
Days of hospitalisation, median (IQR), (range across hospitals)	8 (10) (5–16)		F (6, 514) = 15.89, $p < 0.001$	7 (9) (5–19)	12 (11) (7–14.5)
Comorbidities					
Hypertension	64.3% (40.7–88.5%)	71.82, $p < 0.001$		61.9% (36.4–90.9%)	71.1% (44.4–94.4%)
Other cardiovascular diseases	36.3% (13.6–62.3%)	63.45, $p < 0.001$		36.8% (12.1–70.5%)	34.8% (14.8–70.5%)
Diabetes	29.9% (17.3–54.8%)	34.41, $p < 0.001$		30.6% (16.7–59.1%)	28.2% (0–59.1%)
Chronic obstructive lung disease	15.2% (6.9–32.9%)	40.92, $p < 0.001$		13.2% (0–34.1%)	20.7% (10–44.4%)
Other lung diseases	9% (0–13.6%)	15.03, $p = 0.014$		8.3% (0–18%)	11.1% (0–22.2%)
Neurological disease	7.7% (0–15.1%)	23.14, $p < 0.001$		8% (0–19.7%)	6.7% (0–16.7%)
Mental disorder	7.1% (0–24.6%)	50, $p < 0.001$		6.7% (0–25.6%)	18.2% (0–22.2%)
Liver disease	3.7% (0–17.2%)	34.2, $p < 0.001$		2.6% (0–22.2%)	6.7% (0–15%)
Chronic kidney disease	8.6% (0–22.2%)	41.58, $p < 0.001$		8.8% (1.4–22.2%)	8.2% (0–25%)
Immunocompromised	9.4% (3.3–21%)	16.18, $p = 0.009$		9.3% (4.7–22.2%)	9.6% (0–30%)
Treatment of COVID-19					
Antiviral agents	12.5% (0–11.1%)	71.33, $p < 0.001$		13% (0–11.1%)	11.1% (0–25%)
Corticosteroids	74.3% (34.8–87.9%)	64.25, $p < 0.001$		74.4% (33.3–89%)	74.1% (16.7–100%)
Supplemental oxygen	75.2% (53.1–91.8%)	62.44, $p < 0.001$		69.7% (42.6–88.4%)	91.1% (66.7–100%)

Table 1. *Cont.*

	Total (N = 521) % (Range across Hospitals, %)	Differences between Hospitals		Medical Wards (N = 386) % (Range across Hospitals, %)	ICUs (N = 135) % (Range across Hospitals, %)
		X^2/Fisher's Exact *t*-Test (6, N = 521)	ANOVA		
Number of antibiotics/antifungals					
1	65.6% (51.7–77.8%)			71.2% (58.1–85.2%)	49.6% (33.3–63.6%)
2	26.3% (14.8–32.8%)			25.4% (11.5–41.9%)	28.9% (11.1–50%)
≥3	8.1% (3.5–24.1%)			3.4% (0–11.8%)	21.5% (9.1–35%)
Laboratory findings					
CRP in mg/L, median (IQR), (range across hospitals)	86.7 (107.8) (56–149)		$F(6, 512) = 7.72$, $p < 0.001$	75.8 (97.1) (51.1–118)	120.5 (133.7) (53.2–153)
PCT in µg/L, median (IQR), (range across hospitals)	0.2 (0.4) (0.2–0.4)			0.2 (0.3) (0.1–0.3)	0.2 (0.7) (0.1–0.6)
Leukocyte count in 10^9/L, median (IQR), range across hospitals	7.5 (6) (5.8–11.4)		$F(6, 509) = 13.42$, $p < 0.001$	7 (5.3) (5.6–9.9)	9.9 (8.3) (6.6–10.6)

ICUs: intensive care units; IQR: interquartile range; COVID-19: coronavirus disease 19; CRP: C-reactive protein; PCT: procalcitonin.

Table 2. Antibiotic and antifungal use by type of treatment and indication.

	Total (N = 743)	Medical Wards (N = 510)	ICUs (N = 233)	≤48 h			>48 h		
				Total (N = 425)	Medical Wards (N = 344)	ICUs (N = 81)	Total (N = 317)	Medical Wards (N = 166)	ICUs (N = 151)
Type of treatment									
Prophylactic use; medical	13 (1.8%)	8 (1.6%)	5 (2.2%)	8 (1.9%)	5 (1.5%)	3 (3.7%)	5 (1.6%)	3 (1.8%)	2 (1.3%)
Therapeutic use; empirical	590 (79.4%)	439 (86.1%)	151 (64.8%)	400 (94.1%)	328 (95.4%)	72 (87.7%)	189 (59.6%)	111 (66.9%)	78 (51.7%)
Therapeutic use; targeted	140 (18.8%)	63 (12.4%)	77 (33.1%)	17 (4%)	11 (3.2%)	6 (7.4%)	123 (38.8%)	52 (31.3%)	71 (47%)
Indication									
Pneumonia	564 (75.9%)	379 (74.3%)	185 (79.4%)	374 (88%)	304 (88.4%)	70 (86.4%)	189 (59.6%)	75 (45.2%)	114 (75.5%)
Bloodstream infection	11 (1.5%)	7 (1.4%)	4 (1.7%)	2 (0.5%)	0	2 (2.5%)	9 (2.8%)	7 (4.2%)	2 (1.3%)
Central-line associated bloodstream infection	7 (0.9%)	4 (0.8%)	3 (1.3%)	0	0	0	7 (2.2%)	4 (2.4%)	3 (2%)
Urinary tract infection	41 (5.5%)	37 (7.3%)	4 (1.7%)	14 (3.3%)	14 (4.1%)	0	27 (8.5%)	23 (13.9%)	4 (2.7%)
Skin and soft tissue infection	15 (2%)	15 (2.9%)	0	5 (1%)	5 (1.5%)	0	10 (3.2%)	10 (6%)	0
Intra-abdominal infection	22 (3%)	20 (3.9%)	2 (0.9%)	5 (1%)	5 (1.5%)	0	17 (5.4%)	15 (9%)	2 (1.3%)
Bone and joint infection	7 (0.9%)	4 (0.8%)	3 (1.3%)	0	0	0	7 (2.2%)	4 (2.4%)	3 (2%)
Unknown site of infection	39 (5.3%)	16 (3.1%)	23 (9.9%)	12 (2.8%)	6 (1.7%)	6 (7.4%)	27 (8.5%)	10 (6%)	17 (11.3%)
Other	37 (5%)	28 (5.5%)	9 (3.9%)	13 (3.1%)	10 (2.9%)	3 (3.7%)	24 (7.6%)	18 (10.8%)	6 (4%)

ICUs: intensive care units; ≤48 h: antimicrobials started within 48 h of admission; >48 h: antimicrobials started more than 48 h after admission.

Table 3. Positive microbiology samples and isolated microorganisms.

	Total (*N* = 114)	Medical Wards (*N* = 57)	ICUs (*N* = 57)
Positive Microbiology Samples			
Blood culture	19 (16.7%)	15 (26.3%)	4 (7%)
Sputum	2 (1.8%)	1 (1.8%)	1 (1.8%)
Tracheal aspirate	24 (21.1%)	2 (3.5%)	22 (38.6%)
BAL, mini BAL	23 (20.2%)	1 (1.8%)	22 (38.6%)
Urine culture	28 (17.5%)	23 (40.3%)	5 (8.8%)
Other	18 (15.8%)	15 (26.3%)	3 (5.3%)
Isolated microorganisms			
MSSA	13 (11.4%)	3 (5.3%)	10 (17.5%)
MRSA	3 (2.6%)	1 (1.8%)	2 (3.5%)
CoNS	2 (1.8%)	2 (3.5%)	0
Streptococcus spp.	5 (4.4%)	1 (1.8%)	4 (7%)
Enterococcus faecalis	7 (6.1%)	6 (10.5%)	1 (1.8%)
Enterococcus faecium	5 (4.4%)	2 (3.5%)	3 (5.3%)
Escherichia coli	20 (17.5%)	13 (22.8%)	7 (12.3%)
ESBL-producing *E. coli*	4 (3.5%)	1 (1.8%)	3 (5.3%)
CR *E. coli*	1 (0.9%)	1 (1.8%)	0
Klebsiella spp.	19 (16.7%)	9 (15.8%)	10 (17.5%)
ESBL–producing *Klebsiella* spp.	5 (4.4%)	1 (1.8%)	4 (7%)
CR *Klebsiella* spp.	5 (4.4%)	2 (3.5%)	3 (5.3%)
Proteus mirabilis	10 (8.7%)	6 (10.5%)	4 (7%)
Pseudomonas spp.	12 (10.5%)	5 (8.8%)	7 (12.3%)
CR *Pseudomonas* spp.	1 (0.9%)	0	1 (1.8%)
Acinetobacter spp.	19 (16.7%)	2 (3.5%)	17 (29.8%)
CR *Acinetobacter* spp.	10 (8.7%)	0	10 (17.5%)
Anaerobes	9 (7.9%)	8 (14%)	1 (1.8%)
Clostridioides difficile (toxin positive)	7 (6.1%)	6 (10.5%)	1 (1.8%)
Aspergillus spp.	9 (7.9%)	1 (1.8%)	8 (14%)
Candida spp.	6 (5.3%)	2 (3.5%)	4 (7%)
Other	14 (12.3%)	3 (5.3%)	11 (19.3%)

ICUs: intensive care units; BAL: bronchoalveolar lavage; MSSA: methicillin-susceptible *Staphylococcus aureus*; MRSA: methicillin-resistant *S. aureus*; CoNS: coagulase-negative staphylocci; ESBL: extended-spectrum beta-lactamase; CR: carbapenem-resistant.

The majority of antibiotics prescribed (69.9%) belonged to the "Watch" group of the World Health Organization (WHO) AWaRe classification: cephalosporins (2nd–4th generation), antipseudomonal beta-lactams with beta-lactamase inhibitors, carbapenems, fluoroquinolones, macrolides and vancomycin (Figure 1). In ICUs, antifungals and antibiotics from the "Reserve" group of the AWaRe classification (5th generation cephalosporins, polymyxins, glycylcyclins, oxazolidinones, lipopeptides, etc.) were prescribed more frequently than in medical wards; antifungals accounted for 11.6% and antibiotics from the "Reserve" group for 19.3% of antimicrobial prescriptions in ICUs. The pattern of antimicrobial use varied across hospitals. In some hospitals, antibiotics from the "Access" group (penicillins, beta-lactams with beta-lactamase inhibitors, tetracyclines, trimethoprim with sulfametoxazole, aminoglycosides, metronidazole, etc.) accounted for approximately one-third of prescriptions (including in ICUs), whereas in other hospitals their use was rare.

Analyses of microbiological samples and isolated microorganisms were performed only in patients receiving targeted therapy (Table 3). Targeted treatments with antibiotics and antifungals were based on the results of 114 positive samples. In the medical wards, 40.3% of the positive samples were urine cultures and 26.3% were blood cultures. In the ICUs, bronchoalveolar lavage (BAL) and tracheal aspirate each accounted for 38.6% of the positive samples. The most frequently isolated microorganisms were Gram-negative bacteria: enterobacterales and non-fermenting bacilli. *Escherichia coli* accounted for 22.8%, *Klebsiella* spp. for 15.8% and *Clostridioides difficile* for 10.5% of all isolates in the medical wards. In ICUs, the most common isolates were *Acinetobacter* spp., *Pseudomonas* spp., *Klebsiella* spp., MRSA and *Aspergillus* spp.

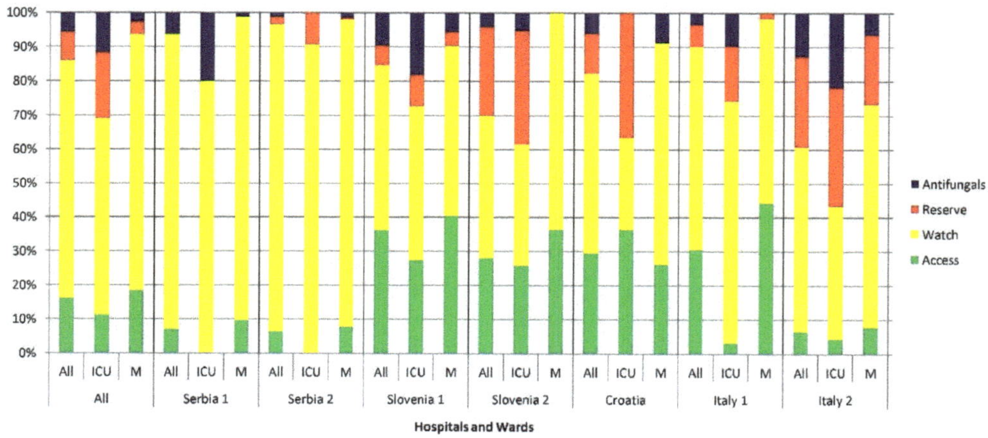

Figure 1. Use of antibiotics (according to AWaRe classification) and antifungals in various hospitals. All: all wards; ICU: intensive care unit; M: medical ward; Access: access antibiotics-penicillins, beta-lactams with beta-lactamase inhibitors, tetracyclines, trimethoprim with sulfametoxazole, aminoglycosides and metronidazole; Watch: watch antibiotics-cephalosporins (2nd–4th generation), antipseudomonal beta-lactams with beta-lactamase inhibitors, carbapenems, fluoroquinolones, macrolides and vancomycin; Reserve: reserve antibiotics-5th generation cephalosporins, polymyxins, glycylcyclins, oxazolidinones and lipopeptides.

3. Discussion

We conducted an international multicentre point-prevalence study to gain information on antimicrobial prescribing in patients with COVID-19 to better guide antimicrobial stewardship in COVID-19 wards. Overall, 52.7% of patients were receiving antibiotics and/or antifungals on the day of the study. The therapy was empiric in most cases, i.e., prescribed on or shortly after admission, with broad-spectrum antibiotics usually used.

The percentage of patients receiving antibiotics and/or antifungals varied between hospitals (32.9–85.6%) and was higher in ICUs than in medical wards (72.6% vs. 48.1%). Reviews examining antimicrobial use in COVID-19 patients reported higher overall antimicrobial use (74.6% and 72% of patients) [7,10]. Prescription appears to vary between different hospitals and countries. In a study conducted in Scottish hospitals, 38.3% of COVID-19 patients were identified as receiving antibiotic therapy [9]; in a study from China 58% were so identified [17]; in some studies from the beginning of the pandemic, the percentage of patients receiving antibiotics was as high as 99% [18,19]. It is worth noting that the period of interest in our study was not the first wave, when uncertainty and common practice were in favour of antibiotic prescribing.

The majority of patients in our study received antimicrobial therapy early in the course of hospitalisation (62% in the first 48 h), but there were significant differences between hospitals: the range between hospitals was 17.4–100%. These results are consistent with other published studies. A study that examined coinfections and early antibiotic therapy in hospitalised COVID-19 patients at four Dutch hospitals found that 60.1% of patients received antibiotics within the first 24 h of admission. Similar to our study, antimicrobial use varied between hospitals: 33.3–72.2% [5]. Another multi-hospital study conducted in the United States of America found early empiric antibiotic use in 56.6% of hospitalised COVID-19 patients, again with wide variation between hospitals (27–84%) [4].

The vast majority of prescriptions was empirical (79.4%), especially early in the hospitalisation. Targeted treatment was more common in ICUs, probably due to the sampling of the lower respiratory tract. The most common indication for therapy with antibiotics and/or antifungals was pneumonia (three-quarters of all prescriptions), especially at the beginning of hospitalisation. Later, it remained the most important indication for antimicrobials, but, especially on medical wards, other hospital-acquired infections, such as urinary tract infections, skin and soft tissue infections and intraabdominal infections, also became an important factor.

The high percentage of antimicrobials stands in stark contrast to the low incidence of coinfections and secondary infections in COVID-19 patients reported in the literature; the overall rate of bacterial or fungal infections is approximately 7–8% [6,7]. Early bacterial coinfections in particular appear to be low, only appearing in approximately 3% of COVID-19 patients, according to studies [4,6,8]. Secondary infections that develop during hospitalisation also appear to be rare, occurring in 4.7–14.3% of patients [6,8].

Broad-spectrum antibiotic therapy was generally used in both medical wards and ICUs. This is consistent with the published literature, in which treatment with cephalosporins, fluoroquinolones, macrolides, beta-lactams with beta-lactamase inhibitors and carbapenems is most commonly described [7,10,18,19]. "Reserve" antibiotics against MDR microorganisms were commonly prescribed in ICUs (19.3% of prescriptions), with resistant bacteria such as *Pseudomonas* spp. and *Acinetobacter* spp. (mainly carbapenem-resistant) accounting for 42.1% of positive ICU samples. MDR bacteria, particularly Gram negative bacilli, are an important cause of hospital-acquired infections in critically ill COVID-19 patients, most commonly of ventilator-associated pneumonia [20,21]. Several outbreaks of infections with MDR bacteria were described during the pandemic waves [14,15,22].

Antifungal drugs were used most frequently in ICUs (11.6% of prescriptions), which was expected due to the fungal infections that occur after the use of high doses of corticosteroids and other immunosuppressive therapies used to treat critically ill COVID-19 patients [23]. In our study, *Aspergillus* spp. was isolated in 14% of positive ICU samples, with 74.1% of ICU patients receiving corticosteroids on the day of the study. One of the first descriptions of an increased rate of pulmonary aspergillosis in COVID-19 patients (CAPA) was performed in Italy, with approximately 30% of patients admitted to ICU diagnosed with probable CAPA, which was associated with high mortality [24]. A study conducted across five ICUs in France showed a lower incidence of secondary fungal infections. A total of 4.8% of patients had a probable/putative invasive pulmonary mould infection, and clinically irrelevant colonisation or false-positive fungal tests were observed in 17.2% of patients [25].

Enterobacterales were most commonly isolated in medical wards (42.9% of positive samples), with urine and blood cultures accounting for 66.6% of positive samples. *C. difficile* was also an important pathogen (10.5% of positive samples), probably due to treatment with broad-spectrum antibiotics. Other studies reported *Streptococcus pneumoniae*, *S. aureus*, *E. coli*, *Haemophilus influenzae* and *Pseudomonas* spp. as common pathogens, which were isolated mainly from respiratory, blood or urine cultures [4,6,8]. In our study, positive respiratory samples were a rarity in medical wards, even though pneumonia was the main indication for antimicrobial therapy. Due to the fact that we did not record negative samples, we cannot distinguish between the low use of respiratory cultures and the low yield of respiratory samples. Studies show that sputum cultures are performed relatively infrequently (7.7–11.4% of patients) due to aerosolisation concerns and the fact that the majority of patients with COVID-19 have a dry, non-productive cough [4,5].

Patterns of antimicrobial use varied among hospitals. Statistically significant differences in patient characteristics show that antimicrobials were prescribed to diverse groups of patients, which may account for differences in prescription. However, various approaches to antimicrobial therapy are more likely to reflect differences in local antimicrobial susceptibilities and antimicrobial treatment guidelines for patients with COVID-19, which are usually in line with local guidelines for pneumonia [26]. There are differences in antimicrobial prescription among the four different countries included in the study. The total consumption of antibacterial agents in DDDs per 1000 inhabitants per day in 2019 was

18.8 in Croatia, 21.7 in Italy, 13.0 in Slovenia and 28.65 in Serbia, with "Watch" category antibiotics accounting for 30% of all antibiotics prescribed in Slovenia and approximately 40% in the other three countries [27,28]. In some hospitals in our study, broad-spectrum antibiotic therapy was initiated on admission, while in others a more prudent approach to antimicrobial therapy was adopted. In accordance with local antimicrobial susceptibilities, hospitals in Serbia and Italy more frequently prescribed antibiotics from the "Watch" and "Reserve" lists of the AWaRe classification. As a consequence of differences in antimicrobial prescribing practices, the impact of COVID-19 on AMR is likely to vary from hospital to hospital, with an increase in AMR expected in hospitals where empirical broad-spectrum antibiotic therapy was frequently prescribed.

The data from our study and the published literature suggest that the use of most broad-spectrum antibiotics is probably unwarranted and unnecessary, especially early in hospitalisation for COVID-19, when the incidence of coinfection is remarkably low. Antimicrobial stewardship teams implementing local guidelines should encourage prescribers to use a more restrained approach. Antibiotics should not be routinely prescribed to patients on admission without evidence of bacterial coinfection. However, it may be difficult to distinguish between severe COVID-19 and bacterial coinfection, especially in critically ill patients. In these patients, microbiological sampling should be encouraged to either guide or discontinue antimicrobial treatment, based on its results. The role of biomarkers such as procalcitonin (PCT) may also be of value. Levels of PCT below 0.1 μg/L have been shown to have a negative predictive value of 98.3% [4]. Other studies have shown that antibiotics can be safely withheld or discontinued in COVID-19 patients with PCT levels below 0.25 μg/L in the absence of other features of bacterial infections [29,30]. The median value of PCT in our study (0.2 μg/L) suggests that many prescribed antimicrobials were probably unnecessary. Higher values of PCT may be more difficult to interpret, because they may have occurred as a result of coinfection or severe COVID-19 [31]. Typical chest radiologic findings in patients with COVID-19 may also help to distinguish between pneumonia due to COVID-19 and bacterial pneumonia [32,33]. Thus, the results of imaging studies are another tool that can help physicians decide whether to treat with antibiotics. As a result of our study, AMS activities were encouraged in the participating hospitals.

Our study provides a comprehensive overview of antimicrobial prescription practices in seven university hospitals caring for COVID-19 patients. It is the first point-prevalence study to examine prescription in medical wards and ICUs in a large cohort of hospitalised COVID-19 patients in several countries and the first report of antimicrobial use during COVID-19 pandemic from the region. Due to its point-prevalence design, it can only provide a snapshot of practices in a particular time window during the pandemic. As information about COVID-19 has grown rapidly, treatment strategies have also evolved over time. Point-prevalence studies can also be influenced by fluctuating prescriptional trends, e.g., biphasic patterns recorded in earlier waves [34]. Due to the fact that information about patient characteristics was collected only for patients receiving antimicrobials, analysis of differences in patient populations was possible only for patients on antimicrobials and not for all hospitalised patients with COVID-19. We analysed only positive samples sent for microbiological diagnostics; therefore, assessment of the frequency of sampling and proportion of positivity was not possible. We also did not specifically investigate the appropriateness of the antimicrobials prescribed, which is another limitation of our study.

4. Materials and Methods

Our multicentre point-prevalence study took place from 11 February 2021 to 15 April 2021 in seven tertiary university hospitals in four countries (Croatia, Italy, Serbia, Slovenia). It was conducted in medical wards and ICUs where patients with confirmed COVID-19 were cared for.

Each COVID-19 ward was audited once and on a single day. All patients 18 years of age and older with confirmed COVID-19 who were hospitalised on the ward at eight o'clock in the morning and were, at the time, receiving systemic antimicrobial therapy were included in the study. The study was conducted by hospital-based physicians. Data were collected about the following drugs: antibiotics for systemic use (Anatomical Therapeutic Chemical (ATC) classification J01), antifungals for systemic use (ATC J02 and D01BA), drugs for the treatment of tuberculosis (ATC J04A), antibiotics for the treatment of intestinal infections (ATC A07AA) and antiparasitic drugs that can be used as antibacterial agents (ATC P01AB).

Two forms were used for data collection (Supplementary Materials, Supplements S1 and S2). The ward form was used to collect information about the number of beds, the number of patients and the number of ventilated patients on the ward (denominator data). Patient forms were used for inpatients receiving antimicrobial therapy (nominator data). For each patient, patient characteristics, days of hospitalisation, days of antimicrobial therapy, indications for antibiotic or antifungal use, type of antimicrobial and its dosage, biomarker levels and radiological findings, use of supplemental oxygen and treatment of COVID-19 were recorded. Antibiotic and antifungal treatment was classified as either prophylactic (medical or surgical) or therapeutic (empirical or targeted). In cases of targeted treatment, microbiological data were also collected. All samples collected for microbiological diagnostics were taken at the discretion of the treating physicians. For ICU patients, information about mechanical ventilation, vasoactive support, pronation and ECMO was also included.

Descriptive statistics were performed using Microsoft Excel. Separate analyses were performed for medical wards and intensive care units and for antimicrobials prescribed within 48 h and after 48 h of hospitalisation. The 2019 WHO AWaRe Classification was used for the analysis of prescribed antibiotics [35]. For analysing differences between hospitals, χ^2 or Fisher's exact test were used for categorical variables and ANOVA for numeric variables. A Bonferroni correction for multiple comparisons was used. Statistical analysis was performed using SPSS Statistics for Windows, version 28.0.

5. Conclusions

In summary, the data from our study show that early empiric use of broad-spectrum antibiotics is common in COVID-19 patients, and that the pattern of antimicrobial use varies from hospital to hospital. The main indication for initiation of antimicrobial therapy is pneumonia. The widespread use of last-line antibiotics identified in some settings, combined with the heavy burden of hospitalised COVID-19 patients, may lead to a substantial increase in AMR. Judicious use of antimicrobials is warranted to prevent an increase in AMR, given the low rates of coinfection and secondary infection reported in the literature. AMS teams should adapt local guidelines accordingly, monitor their implementation and assist treating physicians when real-life dilemmas arise.

Supplementary Materials: The following supporting information can be downloaded at: https://www.mdpi.com/article/10.3390/antibiotics11020176/s1, Supplement S1: Ward form; Supplement S2: Patient form.

Author Contributions: Conceptualization: B.B.; methodology: B.B.; formal analysis: L.P.; investigation: L.P., R.L., B.C., C.T., N.G.M., V.V.P., Z.M.D., O.S., E.S., M.L., G.S., D.P. and B.B.; data curation: L.P.; writing–original draft preparation: L.P.; writing–review and editing, R.L., B.C., C.T., N.G.M., V.V.P., Z.M.D., O.S., E.S., M.L., G.S., D.P. and B.B.; visualization: L.P.; supervision: R.L., B.C., C.T., N.G.M., V.V.P. and B.B.; project administration: L.P. All authors have read and agreed to the published version of the manuscript.

Funding: This research received no external funding.

Institutional Review Board Statement: The study was conducted in accordance with the Declaration of Helsinki and approved by the Ethics Committee of Clinical Centre Rijeka (number 2170-29-02/1-21-2), Friuli Venezia Giulia Ethics Committee (number 0021377), Ethics Committee of University Clinical Centre Srbija (number 111/16), Ethics Committee of University Clinical Centre Kragujevac (number 01/21-141) and Slovenian National Medical Ethics Committee (number 0120-535/2020/5).

Informed Consent Statement: Patient consent was waived due to the non-interventional nature of the study and the fact that all collected data were anonymized.

Data Availability Statement: Available on request.

Acknowledgments: We would like to thank the members of the COVID-PPS study group, Andrej Belačić, Dora Palčevski, Igor Rubinić, Nataša Skočibušić (all Clinical Hospital Center Rijeka, Rijeka, Croatia), Ismet Burekovic, Caterina Di Cecco, Massimo Ferluga, Nicola Mazzuchelli, Mario Santagiuliana (all Azienda sanitaria universitaria Giuliano Isontina, Trieste, Italy), Stefano de Carli, Valentina Gerussi, Alessandro Giancinta (all Azienda sanitaria universitaria Friuli Centrale, Udine, Italy), Matjaž Jereb, Nina Graselli Kmet, Natalija Planinc Strunjaš, Tomaž Vovko and David Zupančič (University Medical Centre Ljubljana, Ljubljana, Slovenia) for their help with organising the study and gathering data.

Conflicts of Interest: The authors declare no conflict of interest.

References

1. World Health Organization. Global Action Plan on Antimicrobial Resistance. Available online: https://www.who.int/publications-detail-redirect/9789241509763 (accessed on 14 December 2021).
2. World Health Organization. WHO Coronavirus (COVID-19) Dashboard. Available online: https://covid19.who.int/ (accessed on 14 December 2021).
3. Hughes, S.; Troise, O.; Donaldson, H.; Mughal, N.; Moore, L.S.P. Bacterial and fungal coinfection among hospitalized patients with COVID-19: A retrospective cohort study in a UK secondary-care setting. *Clin. Microbiol. Infect.* **2020**, *26*, 1395–1399. [CrossRef] [PubMed]
4. Vaughn, V.M.; Gandhi, T.N.; Petty, L.A.; Patel, P.K.; Prescott, H.C.; Malani, A.N.; Ratz, D.; McLaughlin, E.; Chopra, V.; Flanders, S.A. Empiric antibacterial therapy in community-onset bacterial coinfection in patients hospitalized with Coronavirus disease 2019 (COVID-19): A multi-hospital cohort study. *Clin. Infect. Dis.* **2021**, *72*, e533–e541. [CrossRef] [PubMed]
5. Karami, Z.; Knoop, B.T.; Dofferhoff, A.S.M.; Blaauw, M.J.T.; Janssen, N.A.; van Apeldoorn, M.; Kerckhoffs, A.P.M.; van de Maat, J.S.; Hoogerwerf, J.J.; ten Oever, J. Few bacterial co-infections but frequent empiric antibiotic use in the early phase of hospitalized patients with COVID-19: Results from a multicentre retrospective cohort study in The Netherlands. *Infect. Dis.* **2021**, *53*, 102–110. [CrossRef] [PubMed]
6. Garcia-Vidal, C.; Sanjuan, G.; Moreno-García, E.; Puerta-Alcalde, P.; Garcia-Pouton, N.; Chumbita, M.; Fernandez-Pittol, M.; Pitart, C.; Inciarte, A.; Bodro, M.; et al. Incidence of co-infections and superinfections in hospitalized patients with COVID-19: A retrospective cohort study. *Clin. Microbiol. Infect.* **2021**, *27*, 83–88. [CrossRef]
7. Rawson, T.M.; Moore, L.S.P.; Zhu, N.; Ranganathan, N.; Skolimowska, K.; Gilchrist, M.; Satta, G.; Cooke, G.; Holmes, A. Bacterial and fungal coinfection in individuals with coronavirus: A rapid review to support COVID-19 antimicrobial prescribing. *Clin. Infect. Dis.* **2020**, *71*, 2459–2468. [CrossRef]
8. Langford, B.J.; So, M.; Raybardhan, S.; Leung, V.; Westwood, D.; MacFadden, D.R.; Soucy, J.-P.R.; Daneman, N. Bacterial co-infection and secondary infection in patients with COVID-19: A living rapid review and meta-analysis. *Clin. Microbiol. Infect.* **2020**, *26*, 1622–1629. [CrossRef]
9. Seaton, R.A.; Gibbons, C.L.; Cooper, L.; Malcolm, W.; McKinney, R.; Dundas, S.; Griffith, D.; Jeffreys, D.; Hamilton, K.; Choo-Kang, B.; et al. Survey of antibiotic and antifungal prescribing in patients with suspected and confirmed COVID-19 in Scottish hospitals. *J. Infect.* **2020**, *81*, 952–960. [CrossRef]
10. Langford, B.J.; So, M.; Raybardhan, S.; Leung, V.; Soucy, J.-P.R.; Westwood, D.; Daneman, N.; MacFadden, D.R. Antibiotic prescribing in patients with COVID-19: Rapid review and meta-analysis. *Clin. Microbiol. Infect.* **2021**, *27*, 520–531. [CrossRef]
11. Monnet, D.L.; Harbarth, S. Will coronavirus disease (COVID-19) have an impact on antimicrobial resistance? *Euro Surveill.* **2020**, *25*, 2001886. [CrossRef]
12. Kampmeier, S.; Tönnies, H.; Correa-Martinez, C.L.; Mellmann, A.; Schwierzeck, V. A nosocomial cluster of vancomycin resistant enterococci among COVID-19 patients in an intensive care unit. *Antimicrob. Resist. Infect. Control* **2020**, *9*, 154. [CrossRef]
13. Posteraro, B.; Torelli, R.; Vella, A.; Leone, P.M.; De Angelis, G.; De Carolis, E.; Ventura, G.; Sanguinetti, M.; Fantoni, M. Pan-echinocandin-resistant *Candida glabrata* bloodstream infection complicating COVID-19: A fatal case report. *J. Fungi* **2020**, *6*, 163. [CrossRef] [PubMed]
14. Nori, P.; Szymczak, W.; Puius, Y.; Sharma, A.; Cowman, K.; Gialanella, P.; Fleishner, Z.; Corpuz, M.; Torres-Isasiga, J.; Bartash, R.; et al. New Delhi metallo-beta-lactamase producing Enterobacterales infections in New York City COVID-19 patients. *Int. J. Antimicrob. Agents* **2020**, *56*, 106179. [CrossRef] [PubMed]
15. Tiri, B.; Sensi, E.; Marsiliani, V.; Cantarini, M.; Priante, G.; Vernelli, C.; Martella, L.A.; Costantini, M.; Mariottini, A.; Andreani, P.; et al. Antimicrobial stewardship program, COVID-19, and infection control: Spread of carbapenem-resistant *Klebsiella pneumoniae* colonization in ICU COVID-19 patients. What did not work? *J. Clin. Med.* **2020**, *9*, 2744. [CrossRef] [PubMed]
16. Contou, D.; Claudinon, A.; Pajot, O.; Micaëlo, M.; Longuet Flandre, P.; Dubert, M.; Cally, R.; Logre, E.; Fraissé, M.; Mentec, H.; et al. Bacterial and viral co-infections in patients with severe SARS-CoV-2 pneumonia admitted to a French ICU. *Ann. Intensive Care* **2020**, *10*, 119. [CrossRef]
17. Guan, W.-J.; Ni, Z.-Y.; Hu, Y.; Liang, W.-H.; Ou, C.-Q.; He, J.-X.; Liu, L.; Shan, H.; Lei, C.-L.; Hui, D.S.C.; et al. Clinical characteristics of Coronavirus disease 2019 in China. *N. Eng. J. Med.* **2020**, *382*, 1708–1720. [CrossRef]

18. Nori, P.; Cowman, K.; Chen, V.; Bartash, R.; Szymczak, W.; Madaline, T.; Punjabi Katiyar, C.; Jain, R.; Aldrich, M.; Weston, G.; et al. Bacterial and fungal coinfections in COVID-19 patients hospitalized during the New York City pandemic surge. *Infect. Control Hosp. Epidemiol.* **2021**, *42*, 84–88. [CrossRef]
19. Cao, J.; Tu, W.-J.; Cheng, W.; Yu, L.; Liu, Y.-K.; Hu, X.; Liu, Q. Clinical features and short-term outcomes of 102 patients with Coronavirus disease 2019 in Wuhan, China. *Clin. Infect. Dis.* **2020**, *71*, 748–755. [CrossRef]
20. Baiou, A.; Elbuzidi, A.A.; Bakdach, D.; Zaqout, A.; Alarbi, K.M.; Bintaher, A.A.; Ali, M.M.B.; Elarabi, A.M.; Ali, G.A.M.; Daghfal, J.; et al. Clinical characteristics and risk factors for the isolation of multi-drug-resistant Gram-negative bacteria from critically ill patients with COVID-19. *J. Hosp. Infect.* **2021**, *110*, 165–171. [CrossRef]
21. Graselli, G.; Scaravilli, V.; Mangioni, D.; Scudeller, L.; Alagna, L.; Bartoletti, M.; Bellani, G.; Biagioni, E.; Bonfanti, P.; Bottino, N.; et al. Hospital-acquired infections in critically ill patients with COVID-19. *Chest J.* **2021**, *160*, 454–465. [CrossRef]
22. Patel, A.; Emerick, M.; Cabunoc, M.K.; Williams, M.H.; Preas, M.A.; Schrank, G.; Rabinowitz, R.; Luethy, P.; Johnson, J.K.; Leekha, S. Rapid spread and control of multidrug-resistant Gram-negative bacteria in COVID-19 patient care units. *Emerg. Infect. Dis.* **2021**, *27*, 1234–1237. [CrossRef]
23. Machado, M.; Valerio, M.; Álvarez-Uría, A.; Olmedo, M.; Veintimilla, C.; Padilla, B.; De la Villa, S.; Guinea, J.; Escribano, P.; Ruiz-Serrano, M.J.; et al. Invasive pulmonary aspergillosis in the COVID-19 era: An expected new entity *Mycoses* **2021**, *64*, 132–143. [CrossRef] [PubMed]
24. Bartoletti, M.; Pascale, R.; Cricca, M.; Rinaldi, M.; Maccaro, A.; Bussini, L.; Fornaro, G.; Tonetti, T.; Pizzilli, G.; Francalanci, E.; et al. Epidemiology of invasive pulmonary aspergillosis among COVID-19 intubated patients: A prospective study. *Clin. Infect. Dis.* **2021**, *73*, e3606–e3614. [CrossRef] [PubMed]
25. Fekkar, A.; Lampros, A.; Mayaux, J.; Poignon, C.; Demeret, S.; Constantin, J.-M.; Marcelin, A.-G.; Monsel, A.; Luyt, C.-E.; Blaize, M. Occurrence of invasive pulmonary fungal infections in patients with severe COVID-19 admitted to the ICU. *Am. J. Respir. Crit. Care Med.* **2021**, *203*, 307–317. [CrossRef] [PubMed]
26. Beović, B.; Doušak, M.; Ferreira-Coimbra, J.; Nadrah, K.; Rubulotta, F.; Belliato, M.; Berger-Estilita, J.; Ayoade, F.; Rello, J.; Erdem, H. Antibiotic use in patients with COVID-19: A 'snapshot' Infectious Diseases International Research Initiative (ID-IRI) survey. *J. Antimicrob. Chemother.* **2020**, *75*, 3386–3390. [CrossRef] [PubMed]
27. European Centre for Disease Prevention and Control. Antimicrobial Consumption Database (ESAC-Net). Available online: http://www.ecdc.europa.eu./en/antimicrobial-consumption/surveillance-and-disease-data/database (accessed on 14 December 2021).
28. Tomas, A.; Pavlović, N.; Stilinović, N.; Horvat, O.; Paut-Kusturica, M.; Dugandžija, T.; Tomić, Z.; Sabo, A. Increase and change in the pattern of antibiotic use in Serbia (2010–2019). *Antibiotics* **2021**, *10*, 397. [CrossRef]
29. Williams, E.J.; Mair, L.; de Silva, T.I.; Green, D.J.; House, P.; Cawthron, K.; Gillies, C.; Wigfull, J.; Parsons, H.; Partridge, D.G. Evaluation of procalcitonin as a contribution to antimicrobial stewardship in SARS-CoV-2 infection: A retrospective cohort study. *J. Hosp. Infect.* **2021**, *110*, 103–107. [CrossRef]
30. Peters, C.; Williams, K.; Un, E.A.; Little, L.; Saad, A.; Lendrum, K.; Thompson, N.; Weatherley, N.D.; Pegden, A. Use of procalcitonin for antibiotic stewardship in patients with COVID-19: A quality improvement project in a district general hospital. *Clin. Med.* **2021**, *21*, e71–e76. [CrossRef]
31. Zhou, F.; Yu, T.; Du, R.; Fan, G.; Liu, Y.; Liu, Z.; Xiang, J.; Wang, Y.; Song, B.; Gu, X.; et al. Clinical course and risk factors for mortality of adult inpatients with COVID-19 in Wuhan, China: A retrospective cohort study. *Lancet* **2020**, *395*, 1054–1062. [CrossRef]
32. Wong, H.Y.F.; Lam, H.Y.S.; Fong, A.H.-T.; Leung, S.T.; Chin, T.W.-Y.; Lo, C.S.Y.; Lui, M.M.-S.; Lee, J.C.-Y.; Chiu, K.W.-H.; Chung, T.W.-H.; et al. Frequency and distribution of chest radiographic findings in patients positive for COVID-19. *Radiology* **2020**, *296*, E72–E78. [CrossRef]
33. Ye, Z.; Zhang, Y.; Wang, Y.; Huang, Z.; Song, B. Chest CT manifestations of new coronavirus disease 2019 (COVID-19): A pictorial review. *Eur. Radiol.* **2020**, *30*, 4381–4389. [CrossRef]
34. Abelenda-Alonso, G.; Padullés, A.; Rombauts, A.; Gudiol, C.; Pujol, M.; Alvarez-Pouso, C.; Jodar, R.; Carratalà, J. Antibiotic prescription during the COVID-19 pandemic: A biphasic pattern. *Infect. Control Hosp. Epidemiol.* **2020**, *41*, 1371–1372. [CrossRef] [PubMed]
35. World Health Organization. 2019 WHO AWaRe Classification Database of Antibiotics for Evaluation and Monitoring of Use. Available online: https://www.who.int/publications-detail.redirect/WHOEMPIAU2019.11 (accessed on 14 December 2021).

MDPI
St. Alban-Anlage 66
4052 Basel
Switzerland
Tel. +41 61 683 77 34
Fax +41 61 302 89 18
www.mdpi.com

Antibiotics Editorial Office
E-mail: antibiotics@mdpi.com
www.mdpi.com/journal/antibiotics

www.ingramcontent.com/pod-product-compliance
Lightning Source LLC
Chambersburg PA
CBHW062059220926
51850CB00010B/2828
9783036568133